W0036596

The Technological Role of Inward Foreign Direct Investment in Central East Europe

Johannes Stephan

Head, Department for Industrial Organisation and Regulation Economics, Halle Institute for Economic Research, Germany, and Technische Universität Bergakademie Freiberg, Germany (Habilitation thesis)

First published 2013 by
PALGRAVE MACMILLAN

Palgrave Macmillan in the UK is an imprint of Macmillan Publishers Limited, registered in England, company number 785998, of Houndmills, Basingstoke, Hampshire RG21 6XS.

Palgrave Macmillan in the US is a division of St Martin's Press LLC, 175 Fifth Avenue, New York, NY 10010.

Palgrave Macmillan is the global academic imprint of the above companies and has companies and representatives throughout the world.

Palgrave® and Macmillan® are registered trademarks in the United States, the United Kingdom, Europe and other countries.

ISBN: 978–1–137–33375–9

Contents

v

List of Tables

vii

List of Figures

Acknowledgements

I would like to thank all my colleagues who are involved with common research ventures around the IWH FDI Micro Database and the EU projects. Particular thanks go out to the distinguished researchers Wladimir Andreff, Horst Brezinski, Rolf Hasse, and Paul J. J. Welfens for their constructive comments and testimonies provided during my habilitation project. Many other colleagues have provided important comments on the work presented here – my gratitude goes out to all of them.

The publishers and I wish to acknowledge the premission given for the reproduction of copyright material from 'Does Local Technology matter for Foreign Investors in Central and Eastern Europe? – Evidence from the IWH FDI Micro Database' which was published in the Special Issue of the *Journal of East West Business on Market Entry and Operational Strategies of MNEs in Transition and Emerging Economies*, 15/3&4, pp. 210–47 © Taylor & Francis 2009.

List of Abbreviations

CATI	computer assisted telephone interviews
CEE	Central East Europe
CEECs	Central East European Countries
CEFTA	Central European Free Trade Agreement
CMEA	Council for Mutual Economic Assistance
EU	European Union
FA	foreign affiliate of an FDI project
FDI	foreign direct investment
HRSTO	Human Resources in Science and Technology Occupation
IB	international business (literature, theory)
ICT	information and communication technology
IPA	investment promotion agencies
IPR	intellectual property rights
ISCO	International Standard Classification of Occupations
IWH	Halle Institute for Economic Research
MNC	multinational company
MNE	multinational enterprise
NACE	Statistical Classification of Economic Activities in the European Community
NIS	National Innovation Systems
NUTS	Nomenclature of Statistical Territorial Units in Europe
OLI	Ownership, Localisation, and Internalisation Advantages (Dunning's concept)
OLS	ordinary least squares
OPT	outward processing trade
ROR	*Raumordnungsregion*
R&D	research and development
TFP	total factor productivity
TNC	trans-national corporation
TRIPS	trade-related aspects of intellectual property rights
USA	United States of America
WTO	World Trade Organization

1
Introduction

This book intends to assess the role that inward foreign direct investment (FDI) can have on the technological development of the recipient economy, in particular where the recipient country is in the process of economic catching up. This focus positions the analysis in the book between the body of literature testing the existence of economic effects of inward FDI on the one hand, and the management literature concerned with strategies of multinational enterprises in host regions with lower levels of economic development, possibly transition economies, on the other.

The book starts out with a fairly comprehensive introduction that not only discusses the motivation of the topic and the analytical implications that emerge from this focus for empirical analysis, but also provides some of the characterisation of the countries at hand and the issue of FDI itself.

1.1 Motivation of the topic

When the 'historical experiment' of central state planning collapsed in Central East Europe (CEE), it was widely assumed that the formerly socialist countries would be able to swiftly catch up with the West (Lipton and Sachs, 1990; Herr and Westphal, 1991; Sachs, 1992). The Washington consensus (Williamson, 1989, 1993; for critique, see for example Stiglitz, 2002; Rodrik, 2006) explored a path of comprehensive liberalisation of all internal and external spheres of the economy and assumed that entrepreneurial activity,[1] having been subdued for so long, would now thrive in the new pro-economic and competitive environment. Alas, history took a different path: the 'lost decades' of socialist planning had produced technological stagnation and had dried

1

out individual entrepreneurial behaviour (because it was not regarded as a core competence (see Åslund, 2002, p. 283), and would also have run counter to the constituting criterion of the economic plan, resulting in a shortage economy unable to satisfy demand to a very large extent. The liberalisation of the thenceforth 'transition countries' was therefore in need of external help (to overcome the technology gap and to supply demand) – yet help in terms of development aid would have been inconceivable in the light of the large amounts that would have been necessary: the financial 'transfers' invested into the East German new *Länder* are an impressive case in point. Rather, foreign private economic activity had to be the main engine of the transition process and of the process of catching up at least during the first years. This is the role that FDI assumed.

FDI was assigned a pivotal role in the catching-up process of the Central East European countries (CEECs). The list of objectives to which FDI can contribute to at times of systemic transition includes the following, perhaps most prominent, points:[2]

- Alleviating capital shortage: even if doubtful in the medium term (credit gives rise to capital in a monetary economy), this assertion may possibly be acceptable in the case of CEE due to an environment of insufficient trust in the financial spheres around currencies that are newly establishing themselves in terms of international convertibility and acceptance as legal tender.
- Supply of technology: the socialist countries existed in economic autarky from the West within their own integration area of the Council for Mutual Economic Assistance (CMEA), which was both self-inflicted (or governed by the Soviet Union) and at the same time forced upon them by the West, exemplified in the CoCom-list (see for example van Brabant, 1980/2012). They were consequently delinked from the kind of technical advance so characteristic of the development processes in the West since World War II. Those had been broad in the sense of emanating from all sorts of industries and fields of science, were driven to a large extent by private risk-bearing economic activity, and have been – to a large extent, and by consequence – in coherence with demand by the users of new technology.
- Privatisation: who else should have (i) the necessary capital or ability to raise capital to foot the bill of desperately needed recapitalisation and restructuring, and (ii) the management skills, knowledge, and expertise to make the transforming economic entities in those countries internationally competitive?

- Competition: central planning consistently excluded and necessarily had to exclude competition from economic activity. It was foreign trade liberalisation and inward FDI which was to assume the role of enforcing competition on (formerly) state-owned monopolies.

Those four points can be directly deducted from the dichotomy in the world during the socialist era: in economic terms, the distinction between the two worlds, East and West, is best harnessed by characterisation of the constituting criteria of the two systems (see for example Riese, 1991, 1992): whilst the West was governed by competitive and (more or less) free markets that incentivised private economic activity, the East was governed by an overarching economic plan that transcended into nearly all aspects of economic activity. Consequently, economic activity was public and did not involve (or involved to only a very small extent) privately organised economic activities.[3] In the West, the coherence of the economic sphere was based on competition[4] and money. In the East, it was administrative planning. The western world generated an environment conducive to intensified economic foreign integration by way of private foreign investment, private foreign trade and the cross-border migration of labour; the conditions for those were framed in bilateral and multilateral institutions such as the work of the then GATT and the institution of international (reserve-)currency regimes, the International Monetary Fund. The 'communist bloc', on the other hand, planned, controlled, and executed external economic relationships through the political administration such as the CMEA (see, for example, Brezinski, 1978), thereby not allowing the emergence of private internationalising activities (private exporting and importing, outward or inward FDI with internationally owned property rights, international licensing or franchising etc.). Indeed, foreign ownership during socialist times rested by default with the country hosting the property – no other legal arrangement would have been consistent with the system of economic planning. Where new firms were established involving cross-country ownership, international production agreements in intergovernmental contracts guaranteeing 'inter-country socialist ownership' had to align the respective economic plans of respective participants to preserve consistency of the economic plans (see Brezinski, 1978, p. 172–3).[5]

Despite the obvious role that FDI could potentially play in the transition process, the countries in CEE were in reality regarded as unattractive locations for FDI in the early years, due to fears of a return to communism and the significant political risks in some of the countries (for example, the civil war in the former Yugoslavia, the uncertainties souring the break-up of Czechoslovakia, the despotized political regime in the early years of the Slovak Republic, and the non-transparent situation in Romania). When privatisation programmes in the CEE countries started, they flooded the markets with institutionalised productive capital to such an extent and in such a short period of time that prices and values of to-be-privatised firms fell sharply (in Germany to symbolic amounts, and even at times effectively negative due to guarantees and state aid). It was to a large extent foreign capital that took advantage of this (for an early account of the relationship of FDI and privatisation in CEE, see Welfens and Jasinski, 1994, Chapter D). Once the worst transitional recession was overcome in the early 1990s and the economies started on their process of catching up with Western European levels of GDP per capita, CEECs became more interesting targets for FDI. This was spurred not only by mass privatisation, but also by the expectation that these countries would swiftly integrate their already industrialised and now liberalised economies into the European economic area and hence offer profit rates in excess of risk premia. The hope amongst CEECs was that they would not only receive the capital needed to restructure their industries but also that they would get access to modern western knowledge and technology. Most of the CEECs adopted FDI policies based on attracting as much FDI as possible without any concern about the quality of investors (Rugaff, 2008). Over time and with rising stocks of productive capital and levels of economic development, wages also increased. Where wage increases surpass productivity growth,[6] the (unit labour) cost advantages in CEECs melt away; yet now, two decades after the transition process started, CEECs are still offering sizeable unit labour cost advantages in comparison to the West.[7] Where unit labour costs are converging to western levels, strategic motives tend to shift towards building industrial and service networks with the local host economies, reaping localisation advantages beyond motives of pure efficiency. Today, it is the technology-intensive networks that can be expected to be particularly competitive not only as FDI projects, but also as drivers of international competitiveness for the foreign investor in the home country. An increasing number of foreign affiliates in CEECs assume the character of 'developmental subsidiaries', build capacities of technological development, and become active drivers of knowledge and technology in their own right.

This is the empirical starting point of this analysis. The starting point unveils the overarching research question guiding the analysis: under what conditions can inward manufacturing FDI serve as a panacea for transition and catch-up development in CEE?

This analysis cannot, of course, answer that ambiguous question to a satisfactory extent, yet it contributes some important insights that should help generate answers.

1.2 Setting the research agenda: the technological aspect of the 'developmental role of inward manufacturing FDI'

This overarching question forms part of a wider research agenda on the developmental role of FDI in general. The agenda has been evolving for many decades in the bodies of literature on economic development, international economics and, to a lesser extent, in the international business literature. This has produced 'conventional wisdoms' that over time have adapted to new empirical trends and research results generated by more powerful methods and firm-specific data. Three sets of 'conventional wisdom' are summarised in a volume edited by two Washington institutions: the Institute for International Economics and the Center for Global Development (Moran et al., 2005, p. 2):

- The 'Washington consensus' enthusiasm is shared mostly by multinational investors and business lobby groups. It holds that FDI always has a positive effect on the host economy, hence countries should attract as much inward FDI as possible.
- Academic scepticism however prevails over the relationship between inward FDI and economic development. It may well be that the effects of FDI will not yield beyond the effects of any other domestic investment.
- 'Dirigisme resurrected' describes a contemporary tendency by policy-makers in some developing countries to revert to heavy-handed regulation of inward FDI. This may include provisions for minimum technology transfer, compulsory licensing, local content requirements and other measures targeted at increasing technology transfer and spillovers to the benefit of the economy that hosts inward FDI.

This list obviously lacks some more critical views that may not make it to a Washington-accepted 'conventional wisdom' yet still remain important in many host economies, and not only in the third world. Those views direct attention to the possibility that inward FDI may cause environmental damage and may deplete natural resources, and there are allegations related to land grabbing and its potential risks for the indigenous societies. Neglecting those points certainly makes an odd compilation: they may not apply to a large share of inward FDI, they may be overstated in some cases, they may have been more frequent some years or decades ago, and yet they do exist today, and in some cases put immense strain on the host countries (see, for example, the problems associated with environmental damage from oil extraction in some African countries, predominantly Nigeria, see also the critical discussion about the role of recent FDI inflows originating from China around the developing world).

Moran et al. (2005) see the benign effects of FDI to be conditional on and driven by an open trade and investment policy framework that gives rise to competitive markets.[8] On the other hand, where foreign affiliates operate behind trade barriers and without local competition, FDI may even subtract from host country welfare (ibid., p. xii).[9] Today, the empirical literature on average and at the most general level assumes a rather positive view of inward FDI: "FDI is generally associated with positive technological spillovers, economic growth, and increasing income inequality. [...] In all three areas there are, however, significant counter examples in the literature which must be respected." (Clark et al., 2011, p. 2). As much as the Washington consensus can be criticised, it did contribute to generating a competitive and liberalised economic environment for foreign investors, hence it secured the right kind of incentives for efficient economic activity, upon which sustainable economic growth and development can build. It may also be argued that the prospect of EU membership had an even stronger disciplining effect (see perhaps most prominently Lavigne, 1998, in a very large body of literature). It is the securing of such an economic environment that gave rise to the general contention that inward FDI into CEECs had so far positively contributed to systemic change and catch-up development.

This already sets the research agenda for this book: *the objective is to assess the conditions that have to be fulfilled so that inward FDI can have a positive and important role in the technological process of catching up in CEE.* A positive general developmental role, not restricted to the technological impact, may be rooted in a number of effects at very different levels:

- Inward FDI increases employment opportunities for local workers, inasmuch as the additional investment has increased the amount of productive capital which is now in need of a new stock of labour to complement it. The intensity of this effect obviously hinges on the capital or labour intensity of the investment, and on whether employees and managers, in addition to capital, are transferred to the host location. Whilst foreign investment projects tend to pay higher wages than domestic firms (often by trying to acquire the best available human capital embodied in the work force), the best paying jobs are typically reserved for 'expatriates' of the foreign investor, and are not accessible to domestic managers.

- Where inward FDI is associated with capital inflows (depending on the definition of FDI) it increases the volume of capital investment in the host economy – investment that possibly contributes to closing what is called the savings gap[10] in some of the traditional macro-economic development literature (such as the Harrod–Domar growth model, or Rostow's Stages of Growth model, or Nurkse's vicious circle).
- Inward FDI improves the host country's current account via capital imports and thereby contributes to closing the so-called foreign exchange gap. This is obviously a one-off effect for each FDI project, giving rise to no lasting improvements (see the recurring difficulties of, for example, Hungary, where FDI inflow shortfalls keep generating current account problems, or the contagion effects of the current financial crisis, see Brezinski and Stephan, 2011). Moreover, FDI will typically lead to profit repatriation at some stage (if successful, this is after all amongst the most important *raisons d`être* for foreign investment) and may turn this into a negative effect. This has become a relevant issue in Central East Europe, with profit repatriation amounting to around 10 per cent of FDI inflows (in Hungary, this even peaked at slightly over 20 per cent in 2008: Hunya, 2010, p. 13).
- Inward FDI may also have a potentially benign effect on the foreign trade performance of host countries (in particular exports), assuming that foreign firms are more expedient in conquering external markets than are domestic firms (which is often assumed to be the case in emerging markets). This effect, however, depends on transfer-pricing behaviour of foreign investors: where FDI, for example, assumes the character of outward processing trade or if relative corporate taxation rules between countries motivate this, transnational corporations (TNCs) will calculate higher prices for their imports from the internal network and lower values for their exports. Local content rules may reduce this effect where foreign investment is required to generate a minimum share of the value of the products or services in the host country.
- Inward FDI can have effects on competition and market structure in host countries, which is particularly relevant in countries and/ or sectors where an insufficient endowment with domestic firms

competing for market shares results in low intensities of competition. This was in fact an important issue in CEECs. This may equally, however, turn out to be the opposite: if inward FDI involves entry into oligopolistic markets and the foreign firm has more financial leverage than domestic competitors, the foreign firm may rationally try to compete away market shares of domestic firms in a Bertrand competition scenario.

- Inward FDI can also alter sectoral structures,[11] if foreign investment is able to develop successful industrial clusters (which proved to be particularly relevant in Central East Europe – in, for example, the automotive sector in the Czech Republic) and if FDI flows into sectors that were previously underdeveloped but which promise interesting potential (that is, in need of substantial investment, in CEE the ICT industry: Welfens and Borbély, 2009, and Welfens and Wziatek-Kubiak et al., 2005).

- Traditionally, it is also argued that inward FDI will contribute to government revenues of the host country by way of taxation of foreign direct investment and their effects (for example, on employment). Today however, particularly in emerging markets, governments offer generous tax concessions to foreign investors (for example, in special economic zones) and thereby forgo some of this benign effect.

- Inward FDI may transfer knowledge and technology previously unavailable in the domestic economy, if the investment originates from countries with higher levels of economic development and/ or from investors with higher levels of productivity. This knowledge and technology may include anything from hard-core technology to tacit knowledge (for example, technological skills and experience) to management experience and entrepreneurial abilities. It may moreover generate spillovers to the host economy, but only if a number of conditions are fulfilled, such as: if the foreign investor chooses to allow or even promote this in a positive sense, or is unable to prevent the dissipation of its knowledge and technology in a negative sense; if the affiliate of a knowledge-bearing foreign investor is able to convey, teach, and explain the knowledge (even if it assumes a rather tacit character); if the host economy is fit enough to absorb the alien technology and make good use of it; and if the host economy's institutional environment does not put the foreign owner's intellectual property at risk. Whilst this effect depends on many "ifs", it is the technological aspect that is most interesting for the case of inward FDI into CEECs, due to the countries' large technology gaps (see for example Stephan, 2003).

- In addition to technology transfer and spillovers, FDI affiliates may assume an important role as players in the national innovation systems of the host economies. This is particularly relevant in host countries where there is a technology gap between themselves and the home countries of their inward FDI.

- Further, FDI may also have an effect on the distribution of income in the host economy, on labour market structural issues (skills-upgrading, brain drain), on the contribution of FDI to structural/sectoral change (again a particularly important issue in CEECs due to their prior distortion of sectoral structures, see for example Kalotay, 2010). (For comprehensive overviews, see for example Blomström and Globerman and Kokko, 2001; others.)

All the above effects (and possibly some more, not mentioned here) have the potential to increase host economy growth and to spur economic development (see for example Sapienza, 2009, in an application on CEECs). However, each of these beneficial effects may also become detrimental: where the foreign investor is dominant, domestic firms are driven out of the market; foreign investments typically generate more income for domestic groups with lower propensities to save; foreign firms often use their established foreign suppliers (that often follow their clients) and hence drive out domestic firms in upstream industries; FDI projects may crowd out domestic investors seeking credit by raising capital locally on an as yet underdeveloped market; etc.

Adding to these qualifications of potentially benign effects of inward FDI, there are more philosophical or ideological arguments against inward FDI that are nonetheless noteworthy in a discussion of economic effects.[12]

- First, a substantial share of foreign-owned economic activity in large, powerful and dominant firms gives rise to a dualistic economic structure, where the host economy assumes the role of labourer and the foreign investor that of capitalist. Such FDIs will influence host governments, and their lobbying may not always be targeted to the benefit of the local economy. Moreover, dominant FDI projects may divert host economy resources away from 'natural' local comparative advantages (much like the Dutch disease problem). This way, much needed food production is replaced by production of goods catering to the needs of local elites and/or foreign consumers, not the local population at large. Dominant foreign investors may not only drive out domestic indigenous firms, but may also suppress domestic entrepreneurship and thereby inhibit the development of small-scale domestic enterprises.

- Second, foreign investments in emerging markets often seek the localisation advantage of lower labour costs, possibly lower labour unit costs. Where investments are into labour-intensive production technologies, their international competitiveness depends on low wages, and may lock the host economy into a low-wage trap that cannot be overcome without external stimulus. Alas, where investments are into capital-intensive production, the host economy is said to suffer from an aggravated unemployment problem. Finally, where inward FDI is motivated by a laxer protection of labour security, or of the environment or the like, the private profits of the investor may exceed social benefits – and in the worst cases the benefits to the host economy may even turn out to be negative.
- Third, where inward FDI is into countries with less developed infrastructure, it will typically locate in urban centres and hence aggravate the already pressing problems of rural–urban migration.
- Finally, host governments competing for foreign investment projects are able to acquire concessions in the form of excessive (and possibly targeted) protection,[13] tax rebates, investment allowances, and other state aids, either overt or hidden. It is often argued that domestic enterprises do not have this lobby power, because their threat of investing elsewhere is less powerful due to lower investment volumes.

It is obvious that such a discussion about the extent of beneficial and detrimental effects of inward FDI on the host economy is highly subjective and depends to some extent on the individual perception on what kind of economic development is desirable and what character of the development process may on the one hand increase growth but on the other be socially undesirable (see the discussion on the varieties of capitalism, for example Hall and Soskice, 2001; Coates et al. 2005; Cernat, 2006; Tridico, 2011; Myant and Drahokoupil, 2012).

To harness this broad field of interest to something manageable (as well as selecting the issues that are of the greatest interest to the author of this book), the analysis presented here restricts itself to manufacturing industries, because here the largest potentials for technology and knowledge interaction and transfer can be assumed, and because empirical analysis can make use of more robust proxies than would be the case in for example services, where technology is much more about management techniques, and prices and costs are much more difficult to substantiate. The analysis presented here focuses on technology transfer and spillovers as the source of the developmental role via inward manufacturing FDI. This is a controversial issue: considering the conditions that prevail in CEE, will inward FDI in the manufacturing industry assume an active and positive role for technological catch-up development in those countries? How attractive is the region to international investors? What kind of investors does the region attract? Are the manufacturing FDI projects dominated by 'extended work benches', or 'screwdriver industries', or 'outward processing trade'? Do FDI projects originate from investors with the latest knowledge and technology at the frontier area, with its foreign affiliates remaining excluded from this? Do manufacturing FDI projects give rise to modernly equipped production sites in the host economy? And if so, are these fully integrated into the foreign investor's network, that is, isolated from the domestic economy, or do foreign affiliates trade with their host economy? Do affiliates of foreign investors engage in their own technological activities such as R&D and innovation in the host economy? Do they use local firms and institutions in the regional innovation system to exchange knowledge and technology? And finally, which firm- and region-specific factors influence foreign investors' decisions to locate technological activities in a particular region and to source technology locally? Those questions are all at the heart of the technological aspect of the developmental role of manufacturing FDI (in the following, only FDI) and require analysis both at the level of the host economies and at the level of the firm.

The emphasis is on the conditions of such a role and not on constructing a general model or theory of technology transfer and spillovers. Indeed, a model-like description of the mechanisms behind the transfer of knowledge and technology and their spillovers can only be achieved if the concepts of 'knowledge' or 'technology' are treated as a homogeneous goods of groups of homogeneous goods. Whilst this may be applicable to the concept of 'information' to a large extent, knowledge and technology are typically very heterogeneous both in their own character (for example tacit vs tangible and codified, narrative vs formal, trustworthy vs prejudice and misconception, etc.) and in their usefulness for each individual user, which – to complicate matters – may even change over time. Whilst heterogeneities do not in general preclude the development of theories that draw a general picture of a representative average and the mechanisms involved, the many sources of heterogeneity both within the concepts of knowledge and technology and between the many different foreign investors and their host economies make this avenue a very complicated one, to say the least. The analysis here is much less ambiguous, and confines itself to analysing the role of FDI for the 'technological' dimension of the development process by use of empirical firm-level data. The analysis makes use of some of the latest theoretical and conceptual developments in the literature of international business, strategic management, innovation economics, institutional economics, and evolutionary economics.

1.3 Conceptual framework for the analysis

1.3.1 Technological development in economic theories

The 'technological' focus in theories of economic development traditionally operates at the macro-economic sphere and has in particular been less concerned with the sources of technological developments which originate at the firm level.[14] Later additions to the theoretical debates over technology and development integrate the institutional dimensions of international and development economics. Most of these theories are embedded in a world of general equilibrium. Regardless of whether equilibrium is prevalent in reality at any point of time or achieved/restored in the long term, or whether there is such a thing as a tendency towards an equilibrium, those theories still largely remain at variance with what may be considered to be at the root of dynamism in a capitalist market system at the micro-level: that is, uncertainty, transient gains and losses, adjustment costs in friction phases, the stochastic and non-linear character of technical advance, and last but very importantly, the diversity of firm characteristics, reactions, and strategies (firm-heterogeneity, see for an early account Young, 1928). In any dynamic or evolutionary theory, equilibrium cannot exist, because everything keeps changing (see Nelson and Winter, 1982).

Indeed, it may be argued that there is a trade-off between those two families of paradigms: general equilibrium models can be viewed as theoretical benchmarks for assessing the mechanisms at work if the economic system is largely frictionless (perfect competition), and computable general equilibrium models help to assess the likely effects of, for example, policy interventions. Dynamic micro-analyses on the other hand command less sweeping assumptions, hence are closer to the reality of a dynamic world, and yet are rather ill-equipped to predict macro-effects of micro-phenomena (as for example the effects of inward FDI), the outcome of political interventions for the whole economy (even if policies are targeted at the firm level), or exogenous or asymmetric shocks for the whole economy.

Somewhere in this middle ground between the two extremes of macro- and micro-level analyses are meso-level analyses that assess the development of individual sectors (for an application of this on the Central East European ICT industry, see the large body of research on this issue published by Paul J.J. Welfens), or meso-analyses that focus on sectoral change within the macro-sphere, here largely following macro-theories of economic integration, as for example (revealed) comparative advantages in Ricardian or Heckscher-Ohlin frameworks.

1.3.2 The issue of heterogeneity

More recently, other strands of research have tried to assess technological development from the firm perspective, allowing for dynamics at the micro-level and in particular firm-heterogeneity. They are typically listed under the headings 'Innovation Studies', 'Schumpeterian Competition', or 'International Business' and constitute hybrid concepts between (micro-)economics and business studies. However, each of them largely remains limited in scope by focussing on the micro-level with only limited leverage on the macro-level of economic growth, let alone economic development. They also remain largely fragmented in their own theoretical view of the world, lacking a common paradigm.[15]

The work of Melitz (2003) and later developments following for example the 'self-selection' (fixed costs) and 'learning-by-exporting' (endogenous growth models) concepts (see for example Helpman et al., 2004; Helpman, 2006, in a world of general equilibrium) may serve as a link to incorporating firm-heterogeneity into technological development by modelling the impact of trade liberalisation (and a reduction of other institutional trade costs) on firm productivity – and via inter-firm reallocation on aggregate productivity of technological development (generating Ricardian comparative advantages, see for example Love and Mansury, 2009).[16] Another possible path of theory development may be derived from micro-level analyses, namely CDM models (Crépon, Duguet, and Mairesse), linking innovation inputs at the micro-level to innovation outputs and labour productivity at the firm level. But how do the insights generated in these models link up with the macro-problem of economic development?

The most straightforward way to link micro-economic studies that are able to consider heterogeneity for the issue of technological development and the macro-level of economic catch-up development is to conceptualise the issue of heterogeneity as a micro-foundation and simply aggregate as usual, but by use of groups of firms and respective economic sectors that share common characteristics with respect to the technological effects. This, however, is still far off the kind of heterogeneity observed in FDI issues: for example in CEECs, FDI projects in high-tech sectors sometimes remain even less technologically active than FDI projects in more traditional industries: FDI may be used to conquer markets, with the local foreign affiliates having no part in the technological development of for example pharmaceuticals; or: even high-tech industries will require some degree of low-tech activities – being intellectual property-sensitive, firms that want to exploit potential efficiency gains from investing in CEECs may prefer to organise these activities under ownership control rather than outsourcing (the internalisation-issue).

Another possibility to link micro-technological development with macro-economic technical advance is by assessing the role of institutional framework conditions for the technological activity of firms: it is institutions that give rise to a framework that incentivises a particular behaviour. The effects of these incentives may vary in magnitude and intensity across firms and other actors of an innovation system, but the direction of those effects can be assumed to be rather homogeneous across heterogeneous firms in heterogeneous host countries.

Today, equilibrium models often are held to remain important as theoretical reference points for analysis explicitly departing from the equilibrium paradigm: this is because the former are based on pure deductive model-building, whilst the latter are less strict in the sense of allowing even conflicting explanations and predictions. It may hence be argued that strict analysis still has to consider a stringent theory, even if the matter of analysis is in direct opposition to the assumptions of the theoretical reference point. For firm-level analysis, this may for example involve the extension of analysis of technological effects of FDI that are rooted at the level of the firm to the implications of these micro-effects for technical advance at the macro-sphere. And this would have to be done within the framework of general equilibrium models: where, for example the analysis finds positive technology spillovers emanating from foreign firms in a particular sector (that is, positive technology effects at the micro-level), the interpretation of their macro-implications would still require consideration of implications on foreign trade specialisation, of effects on capital/labour ratios, of effects on labour markets, on aggregate demand, wages, savings, etc. Unfortunately, the results of such kinds of analysis remained highly ambiguous, even where very powerful analytical techniques were applied.

The analysis presented here, however, follows a different path: the conditions of positive technological effects at the firm level are deducted from concepts and theories at the firm level and tested empirically for the particular case of CEECs. By identifying the relevant conditions, this use of theory is able to make a convincing proposition for the technological role of FDI at the level of the firm. Whilst this does not necessarily pertain to effects on the level of the whole economy with interactions between firms and sectors and markets and macro-aggregates, whilst the conditions for the translation of these technological effects to aggregate total factor productivity remain unclear, this conceptual framework does single out the pre-conditions for technical advance at the macro-level that emanates from the channels generated by inward FDI at the micro-level.

1.3.3 The method of analysing conditions of technology transfer and spillovers

A large share of the literature concerned with the technological aspect of the developmental effect of FDI aims to measure the amount of technology transfer and spillovers. It often uses firm-level AK-models (production functions, TFP analysis of firms or of firm groups). Despite its large size, the results of this body of literature still remain largely inconclusive, which may be owing to the large extent of heterogeneity in the firms as subjects of analysis, as well as heterogeneity of foreign investors, the host and home economies, etc. Lipsey and Sjöholm (2005) hence hold that "the search for universal relationships [with respect to the impact of FDI on home countries] is futile" (ibid., p. 40).

Industry-, region-, or country-specific analysis of extents of knowledge and technology transfer or spillovers is not conducted here, because of the many measurement problems and the resulting inconclusive results generated in the past on this issue. With respect to industries, it has to be noted additionally that the statistically usual classifications typically do not sufficiently represent homogeneous technologies applied at the shop floor: even high-tech industries (see for example the classification by Pavitt or the OECD of the one from Vienna) involve some rather low-tech tasks, and often FDI splits tasks between regions according to the intensity of knowledge and technology. So even if FDI in CEECs is found in high-tech industries, the technological content of tasks in the foreign affiliates in the East may still be rather limited. The defining features of TNCs include their ability to shop around in their locations for differences in production or indeed in all factors, and hence they will tend to allocate different tasks to different locations according to the advantages of those locations. The life-cycle theory of Hirsch (1965, 1967) puts this idea into a timeline, suggesting that each location has an advantage according to the life-cycle stage of the product or industry. Additionally, this pattern of strategic allocation of tasks between different locations by the foreign investor may be a result of different strengths of intellectual property rights regimes. Whatever the dominant reason, all this suggests that the conditions for technology transfer do not become much more homogeneous even if analysed in narrowly defined industries.

Furthermore, this analysis does not envisage the development of a formal model, due to the problems associated with the heterogeneity issue. Rather, this analysis attempts to identify conditions necessary for the process of technological catching up via inward FDI. Those could then feed into a dynamic theory of economic development via inward FDI (without being able to provide a complete, sufficient list). The research objective is to derive from theoretical literature assumptions on what determinants can be expected to increase the extent of technology transfer and spillovers, and to develop and test hypotheses on firm-specific, country/location-specific, and sector-specific conditions of such determinants for the particular case of CEECs. Such a method not only generates new knowledge about specific conditions of (determinants assumed to give rise to) intense technology transfer and spillovers in this region, but, moreover, the assessment of the situation within countries, regions, or industrial sectors allows the estimation of potentials to benefit from technology transfer and spillovers via inward FDI. Finally, some of these conditions may lend themselves as instruments for active technology transfer policy.

Examples of firm-specific conditions include: internationalisation of R&D and innovation; strategic investment motives; its own technological activity of affiliates; absorptive capacities or capabilities of affiliates; corporate governance between investor and affiliate; and integration and embeddedness within the local host economy (see Cantwell's internal networks). Examples of host region-specific conditions include the size of demand and the market in general; agglomeration advantages such as the infrastructure and clusters; the availability of qualified personnel; capabilities of all actors in the national/regional/sectoral innovation systems such as supplier and consumer industries, competitors, user industries, other networking partners, public and private research organisations, local and regional public administration, and policy-makers (see Cantwell's external networks); institutional framework conditions such as the rule of law, IPR regime, competition law and policy; and technology transfer policy such as innovation systems and policies, local content rules, etc.

In positive methodological terms, the central hypothesis guiding the research is that the prospects for the catching-up development of an economy depend firstly on its being endowed with certain specific characteristics of its firms and industries, including their ability and actual capacity to adopt, adapt and change, that is, their technological capabilities. Secondly, they depend on the state of development of institutions and the fit between formal and informal, between existing and development-friendly institutions. Thirdly, they depend on a policy mix that takes account of both the set of necessary conditions for technological catch-up and the interactions between policy instruments and their effects.

1.3.4 Connotations and definitions

Because the research of this book is based on several bodies of literature (international economics, development economics, international business, international management, innovation economics, etc.), some terms may have different connotations attached to them, depending on where they emanate from.

The concept of 'economic catch-up development' is defined as a dynamic process[17] in which the economy reaches higher levels of aggregate income per capita. This is clearly different to the concept of 'economic growth', which does not distinguish between the growth of quantitative factors of production and the growth of national value added without increases in either capital or labour. The latter concept is based on aggregate productivity growth and is a result of improvements in the use of scarce resources or production factors, which may be subsumed as technological progress (see the typical growth theory-discussion). The catch-up concept is also dynamic at the micro-level, as it entails that "economic actors take action that breaks away from the kind of behaviours previously relied upon" (Nelson, 2008, p. 15). If economic agents are able to thrive in this ever-changing environment, they may gradually achieve higher levels of income through improvements in the technology that underlies economic activity. Technological advance may originate from improvements in the technology used in economic activity and/or from new and improved products and services. The connotation of catching up further narrows the meaning of the term 'economic development' inasmuch as it assumes the character of catching up to levels of economic development in more advanced countries by way of 'closing the gap' between the technologies used in economic activity, and/or by way of adopting the superior products and services in the more advanced countries, and/or by way of structural change towards sectors with higher productivity levels. Hence, whilst a strict interpretation of this connotation of catching up does not include the possibility of the catching-up economy generating new and improved products and services as yet unavailable in the more advanced countries (giving rise to a hierarchy of countries), this may in reality still occur, and this possibility should be included in the connotation of catch-up development.

'Foreign direct investment' is a sufficiently general term to be quite unambiguous: FDI describes the kind of internationalisation that involves the control of an investment by way of ownership in a firm located in a different jurisdiction from that of the original home location of the investor. It is often difficult to establish the original location of the headquarters; some internationalised firms even have several headquarters, depending on the functions they are endowed with. Also, it is often not worthwhile going back as far as the original owner of a firm, as the home location of the ultimate owner may not have any effect on their investment in another country, where the firm controlling its investment in a foreign country resides in a country other than that of the ultimate owner. Hence in this analysis the home country of the FDI is the country of residence of the investor who controls the foreign investment. For statistical purposes, the minimum extent of ownership that is required to make an investment in a foreign country a 'direct' investment project is often set at 10 per cent ownership share (World Bank, UNCTAD). The assumption is that a 10 per cent share is associated with a long-term engagement, whereas short-term investments are rather termed 'portfolio investments'.

Firms that have internationalised into more than two foreign locations are often called 'multinational co-operations' or 'multinational companies' (MNCs). Because this analysis does not distinguish between firms engaged in two or more foreign locations, the term 'transnational corporations' (TNCs) is used when discussing their own analysis (where literature is reviewed, the original terms in the literature are used).

In the different strands of literature, several terms are used for the firms that form part of an internationalised firm: as well as the headquarters firm, a TNC may consist of subsidiaries, foreign-invested firms, foreign-owned firms, or foreign affiliates. The connotations of those terms are not always homogeneous, and depend on issues such as the ownership share, the legal status, and the number of different locations in which the TNC or its owner's network is internationalised. Probably the least demanding term with regard to such characteristics is 'foreign affiliate', which includes all firms that are part of an FDI project in some form or other. Because the firms in the data used in this analysis comprise all kinds of firms with foreign ownership, the term 'foreign affiliates' is used when describing my own analysis (again, where literature is reviewed, the original terms in the literature are used).

The movement/diffusion of knowledge and technology from one potential or actual user or bearer of knowledge and technology to another, may, in the world of internationalised firms, include an intentional transfer of tangible technology (for example patents, business secrets), the intentional transfer of tacit knowledge (for example know-how, experience made available by way of training of foreign affiliate personnel by personnel of other members of the network), the intentional transfer of knowledge and technology embodied in products or services, or the unintentional dissipation of embodied knowledge and technology, and finally the unintentional dissipation of knowledge and technology via imitation, learning, etc. Often, the relevant literature distinguishes between technology transfer and spillovers in clearly distinct connotations: spillovers are externalities, not peculiarly remunerated, whereas technology transfer denotes the intentional diffusion of knowledge from one individual (employee) or institution (firm) to the other. Blomström and Kokko (2003) differentiate between internal effects of inward FDI (giving rise to direct transfer of technology between the foreign investor and its affiliate) and external effects (between the foreign affiliate and domestic firms, the traditional spillover effects). The external indirect transfer of technology from the foreign affiliate to other host economy firms is further split up into vertical transfer along the supply chain (to upstream suppliers via backward linkages, or to downstream customers through forward linkages) and horizontal transfer from the foreign affiliate to host country firms in the same industry. Complicating matters, however, direct internal diffusion of knowledge between foreign investors and foreign affiliates may be intentional (technology transfer) or unintentional and not accounted for in the contractual arrangements between the two sides (spillovers/externalities). The same applies to the movement of knowledge between the foreign affiliate and firms outside the foreign investor's network. At times, this distinction may be relevant for a precise conceptualisation of analysis; however, it is implicitly assumed in the relevant literature that internal technology diffusion probably never constitutes externalities, because it happens within the same firm (even if between different locations). This is in fact a pragmatic solution, inasmuch as empirically, the distinction between transfer and spillovers in the internal sphere within a firm is very difficult if not impossible, not least due to 'transfer pricing'. In this analysis, the term 'internal technology transfer' is used regardless of whether the diffusion of knowledge and technology is intentional or not, and 'internal spillovers' giving rise to externalities are subsumed under the heading of 'internal technology transfer'. For the external diffusion of knowledge and technology, this analysis also aligns itself with the relevant literature, distinguishing between 'transfer' (intentional, contractual, no externalities) and 'spillover' (may be unintentional, is not regulated in contracts, hence gives rise to externalities).

In the conceptualisation of the long-term research programme, the distinction between external technology transfer and external spillovers does not play a role. This generalisation is mainly due to difficulties in empirically distinguishing the two and was successfully used throughout many projects, including in two Fifth and Sixth EU Framework Programme research projects (PRODGAP[18] and UKNOW[19]) that gave rise to the still evolving IWH FDI Micro Database. The research programme rather focuses on the technological role of FDI with their foreign affiliates in their host economies. Here, the programme distinguishes between (i) the role of the host economy (conceptualised as the 'host innovation system' with its main actors of local suppliers, local consumers, and local scientific research institutions) for R&D and innovation activities in the foreign affiliate, and (ii) the role of the foreign affiliate for technological activity in the host innovation system (that is, the object of interest of the spillover literature). In contrast to much of the spillover literature, this research programme does not measure the extent of technology transfer and spillovers (whether internal or external), but rather deduces from previous theoretical and empirical work what determines a positive technological role of FDI in host economies and of host economies in foreign affiliates, and tests those on the particular case of CEECs.

1.3.5   The selection of countries

The focus on technology for economic development is most relevant for the formally centrally planned economies in CEE: they provide specific insights into the study of the conditions of technological catch-up development economies. This pertains not only to the time since the onset of systemic transition, but in particular to it: first and foremost, the vast privatisation programmes for the formerly large monopolist state-owned companies as well as the importance of FDI for these privatisation programmes not only shaped industry structures in and during transition, but are also an important explanatory factor for today's market structures. A wide range of different ownership structures and corporate control mechanisms emerged from the different privatisation programmes (Claessens and Djankov, 1998): small and medium-sized companies were typically privatised by means of a variety of different methods; the majority of large firms in most countries were privatised to strategic investors, with a crucial role played by foreign investors (see Major, 2003; Mickiewicz and Baltowski, 2003). This makes an analysis of technological development via FDI a very incisive and instructive field of study. It further suggests that competition policy would have to be tuned to national particularities to positively act as an engine for technological development (mainly concentrated markets, often dominated by foreign direct investors, with state aid often attuned to the needs of FDI and privatised firms).

Moreover, those economies still have a relatively recent industrial history before they embarked on their socialist adventures, and they have experienced a complete overhaul of institutions of economic governance after the overthrow of the ailing system. After having been largely *de*linked from the West for some decades by way of economic autarky from the West (self-inflicted and forced upon them by the West, exemplified in the CoCom-list, see for example Welfens, 1995, pp. 12–13), foreign and domestically generated technology falls on fruitful soil today in terms of acceptance of new technology and technology-governing institutions (even if the latter are often still underdeveloped, and not as yet fully transformed).

During socialist times, only a handful of the formerly state-owned companies had a licence to export, let alone to internationalise by way of outward foreign direct investment, so that today it is mainly inward foreign direct investors who drive the internationalisation processes of the transition countries. In contrast to most (other) emerging markets, institutions in CEE are today well aligned to an industrial society based on division of labour and on the uncertainties and risks involved in a dynamic and competitive economic system (for example information asymmetries, tacit knowledge, etc.).

Finally, the economies in Central East Europe are intensively integrated with Western Europe, not only giving them access to large markets, the capital needed for investment, and advanced foreign technology, but also making them prime locations for FDI from the nearby West with its anticipated potential positive effects on technology transfer and spillovers.

The Czech Republic has declared the transition process of its economy to be completed; the European Bank for Reconstruction and Development, the World Bank subsidiary specialising in CEE, rates the most advanced CEECs (in terms of 'transition indicators') at levels close to or even at par with levels in the so-called market economies as a benchmark; the

European Union has likewise established in its 'coherence reports' that the most advanced CEECs – namely. the three Baltic countries, Poland, the Czech and Slovak Republics, Hungary, and Slovenia – can now be considered fully-fledged 'market economies'. Hence, most relevant institutions believe that these countries have mastered their monumental task of transition from a state-controlled, socialist or communist system of economic governance to a system governed by the mechanisms of competitive and open markets.

The most advanced CEECs having been full members of the European Union since March 2004, do they offer their entrepreneurs framework conditions for economic activity that correspond to the kind of competitive environment with its opportunities and constraints that prevail in Western Europe and in the United States of America – that is, typical market economies? Are the market mechanisms that govern economic activity in those economies, and the institutions that influence economic activity in market economies, (more or less) uniform in comparison to those in the established market economies?

During the burdensome transition process, with recessions and a profound overhaul of complete economies, whilst most of the important and difficult reform steps have been undertaken and formal institutions installed in these countries, whilst economic opportunities and activities in those countries today appear to fulfil the criteria of market-type economies, strong differences still prevail between East and West Europe, painting a picture rather different from that proclaimed by the most relevant institutions accompanying the transition process: (i) some important formal institutions (laws and regulations) are not yet sufficiently in line with informal institutions (norms and values of economic actors), as revealed by a comparison of *de jure* and *de facto* rules of the game in economic activity; (ii) entrepreneurial opportunities in those economies are clearly constrained in ways other than in the West, in particular with respect to the competitive environment for economic activity.

Arguably, differences will always prevail between even the most advanced market-based economies. In the cases of the CEECs, however, some commonalities in their economic features distinguish them from the West, and those differences are reminiscent of their socialist past – transition may be over, but transition-related particularities remain in place.

Those differences are in fact of paramount importance for the task ahead in all transition countries in CEE – the task of catching up, in terms of levels of economic development in general and in terms of income per capita in particular. Because these economies, as distinct from other emerging markets or developing countries, are emerging from an industrialised past (even if driven and governed by diametrically different objectives and institutions), their people are well accustomed to an industrial way of life of division of labour, a hard budget constraint with respect to earnings and ability to afford consumption (yet importantly: not investment). Furthermore, the countries benefit from a well educated population and their economies from a well trained workforce. **Hence the decisive criterion for economic catching up is technological development:** whereas in emerging markets such as China or in the developing countries of sub-Saharan Africa economic development is based both on technical advance and on a sheer increase in the number of employed production factors (extensive growth), in CEE catching up is much more constrained to improvements in the productivity of use of production factors (intensive growth).

1.4 Empirical evidence of inward FDI into Central East Europe

First and foremost, the potential technological role of FDI into any region depends on the significance in terms of size of FDI in the host region. CEECs have in fact attracted large amounts of FDI since their opening up to the rest of the world, even if their global shares of FDI inflows still are well below those of most developed and many emerging markets.

All countries in East Europe have received a considerable amount of inward FDI in the most recent past, the vast majority of which originated from Western Europe (Hunya, 2010, p. 41)[20]. It is particularly important to note that the CEECs had practically no inward FDI before 1990. Since then, FDI with better prospects in terms of a technological role for their host countries has been replacing the early tendency to attract outward processing trade (Andreff and Andreff, 2001). Altomonte and Guagliano (2003) report that CEECs had a greater potential to attract FDI than the Mediterranean region. The resulting importance of FDI in their host countries in CEECs effected a strong intensity of real economy integration between East and West Europe, as well as their integration into the global (innovation) networks of their western foreign investors.

The following displays a large amount of descriptive statistics about inward and outward investment, home countries and their sectoral distribution. In the first part, a comparison is made between the role of FDI in CEECs with that in other emerging markets, namely the

BRICs: Brazil, Russia, India, and China, plus Germany as an example of a developed country (the eastern region of which, however, is also a post-socialist economic region). The second part uses largely the same indicators, and makes comparisons across the ten CEECs that are today members of the EU.

1.4.1 Comparison BRIC and aggregate CEE-10

FDI into CEE-10 plays a rather small role on a global scale when compared to other emerging markets (BRICs). Whilst displaying a considerable degree of heterogeneity, FDI inflows into BRICs accumulated higher

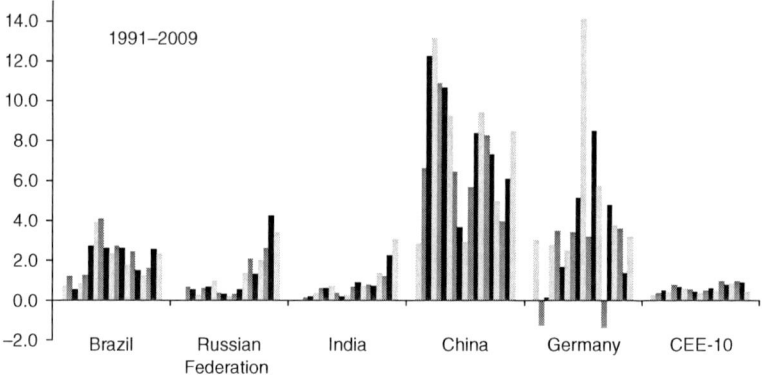

Figure 1.1 FDI inflow, in % of World FDI inflows
Source: WIR, 2010, FDI database.

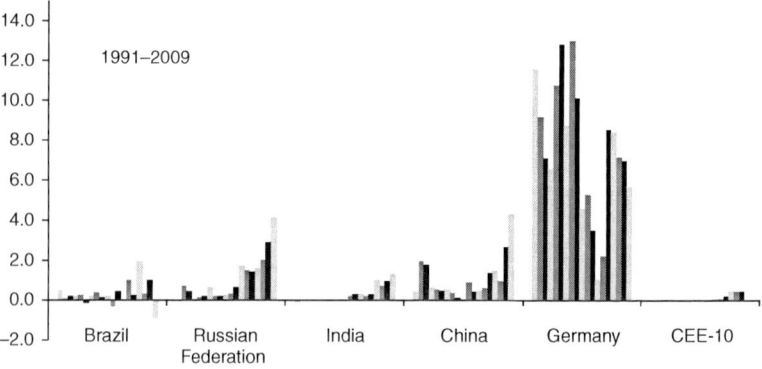

Figure 1.2 FDI outflow, in % of World FDI inflows
Source: WIR, 2010, FDI database.

world shares in most years. China accounts for exorbitantly large shares, not least due to its size, whilst Germany, here exemplifying the case of a developed country, is somewhere between China and emerging markets (the substantial jump in Germany in 2000 is an outlier and, with the acquisition of Mannesmann by VodafoneAirTouch, represents the largest ever deal worldwide).

Of even less significance is the share of FDI outflows from CEE-10 within total world FDI flows. Whilst the tendency of emerging markets to remain less active in terms of FDI outflows is also mirrored by the BRICs, the rates of growth and world shares in the latest recorded years suggest rapid catching up in this aspect. It is mainly (oil-rich) Russia and (export-rich) China that are exhibiting such impressive rates of growth; both are noted for the wealth they have accumulated in their sovereign funds in the past decade. China is prominent as the country that receives much larger shares of World FDI inflows than can be observed in any other emerging markets yet remains within this league for outward FDI. This very much reflects the fact that outward FDI is still dominated by the developed world whilst emerging markets have become particularly interesting for inward FDI. Germany, as would be expected for a developed country, captures much larger shares in world outward FDI than does any emerging market, and also larger shares of outward than inward FDI.

The importance of inflowing FDI for the CEE-10 economies becomes apparent if annual inflows are corrected by GDP levels: here, FDI in CEE-10 has become more important than in any of the BRICS or

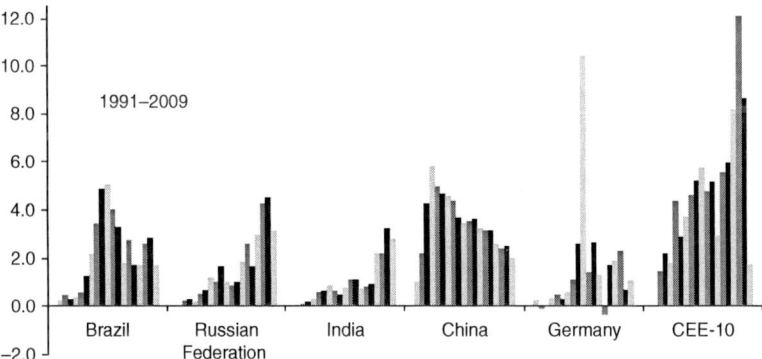

Figure 1.3 FDI inflow, in % of GDP
Source: WIR, 2010, FDI database.

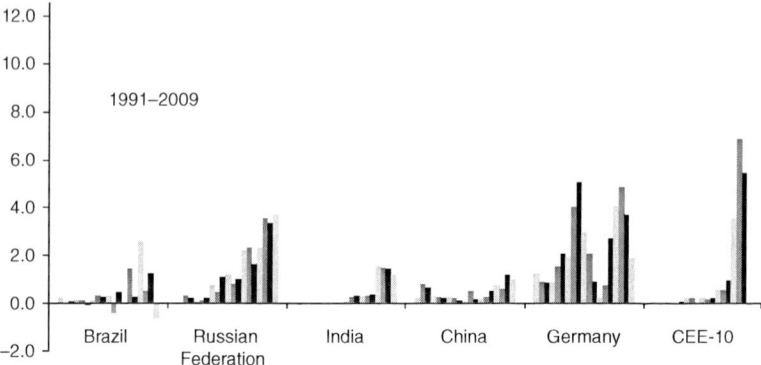

Figure 1.4 FDI outflow, in % of GDP
Source: WIR, 2010, FDI database.

Germany. On the other hand, inward shares in national product have markedly declined in China and to some degree also in Brazil. It is also important to note that the CEE-10 region has suffered significantly in the course of the current financial crisis, that is in 2008 and 2009; the other countries have also seen an FDI inflow setback, but certainly not as strong as that experienced by the CEE-10.

Inward FDI has always dominated outward FDI in emerging markets. The recent growth of outward FDI from emerging and transition economies may be assumed to be an effect of the expansion of ICT and the acceleration of innovation dynamics associated with this (see for example Welfens, 2011, p. 135). Outflowing FDI has been much less important for the economies of the 10 CEECs group during the early transition phase, but has become increasingly more important in the years 2006 and 2007 – and, alas, the financial crisis has also put a hold on outflows in 2008. The data for 2009 is not yet available but can be expected to be even lower than for 2008.

The immense importance of inflowing FDI for the economies in CEE-10 is probably most obvious in its effects on national investment: the share of foreign direct investment in total national gross fixed capital formation has become considerably higher than in any other of the comparator countries ever since 2006. Even in Brazil, where at peak periods some 30 per cent of this type of investment was covered by foreign sources, CEE-10 had on average a rather steady growth starting from less than 10 per cent in the early 2000s to a one-off maximum so far of over 50 per cent.

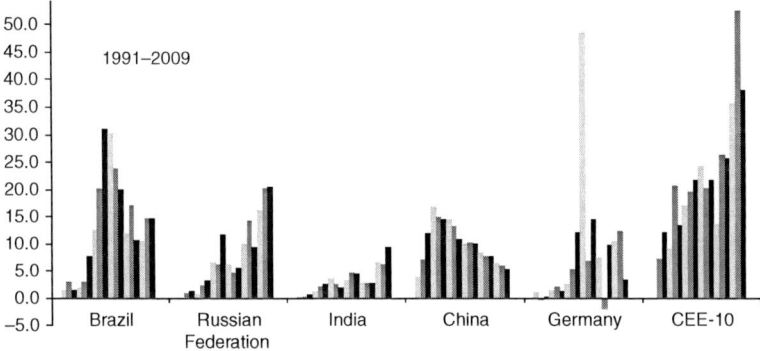

Figure 1.5 FDI inflow, in % of Gross Fixed Capital Formation
Source: WIR, 2010, FDI database.

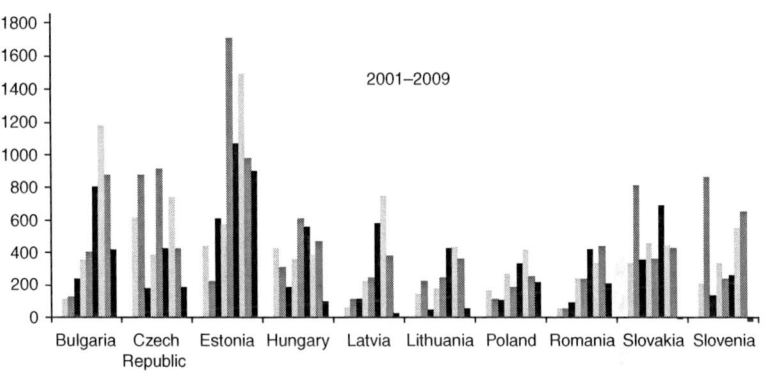

Figure 1.6 FDI inflow, in EUR per capita
Source: wiiw, 2010, p. 24.

1.4.2 Comparison between individual Central East European countries

FDI inflows into CEE were still on the rise until the mid 2000s, and here in particular in CEECs that are less developed in terms of GDP per capita and/or in terms of progress with systemic transition (measured in the process of restructuring of the economy and institutional development).

However, the recent financial crisis effected sharp capital inflow setbacks, the size and timing of which are quite heterogeneous between the countries: most CEECs experienced setbacks in 2009, with the exceptions of Estonia and Poland in 2008, Bulgaria, the Czech Republic and Latvia during both 2008 and 2009, and Hungary and Slovakia in 2007.

Until the mid-1990s, it was the Czech Republic, Hungary, and Poland (that is, Visegrád without the Slovak Republic) which had been the main recipients of inward FDI amongst CEECs. More recently, Bulgaria, Romania, Slovakia, and Slovenia have caught up considerably, whilst Poland has outpaced all other CEECs. Inward FDI stocks in the three Baltic countries remain comparatively low in absolute terms.

Where foreign investors engage in a host country with whatever strategic motive (market access, following a key client, efficiency-seeking investment motives, access to localised knowledge and technology, and access to natural resources), they expect some return on investment in the end. Their propensity to be able to achieve this is measured in Figure 1.8 in terms of FDI-related income outflows; it measures the share of pecuniary profits generated by FDI that are not reinvested (either in the host economy or used for further FDI into third countries) but are repatriated to the home country as income.

This then constitutes the 'loss' or opportunity costs that a country incurs due to the fact that investors are foreign. In CEE, FDI now has a history of up to slightly more than two decades, so profit repatriation plays a role; it currently amounts to some 7 to 12 per cent on average. In Hungary, Estonia, and the Czech Republic, income repatriation is the highest, but in the three Baltics and Romania the current financial crisis has effected a large fall in repatriated FDI income (with negative rates per inward FDI stocks in Latvia and Lithuania).

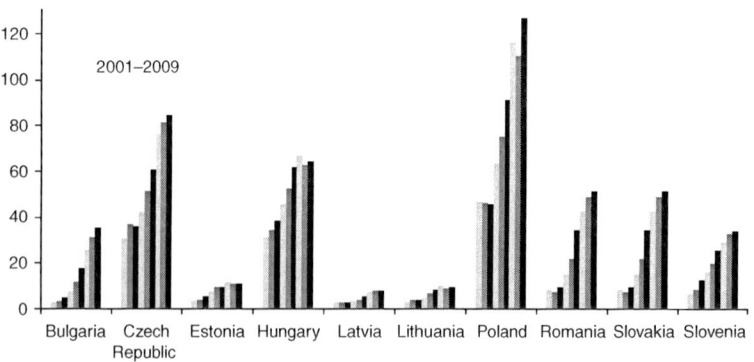

Figure 1.7   Inward FDI stock, in EUR mn
Source: wiiw, 2010, p. 24.

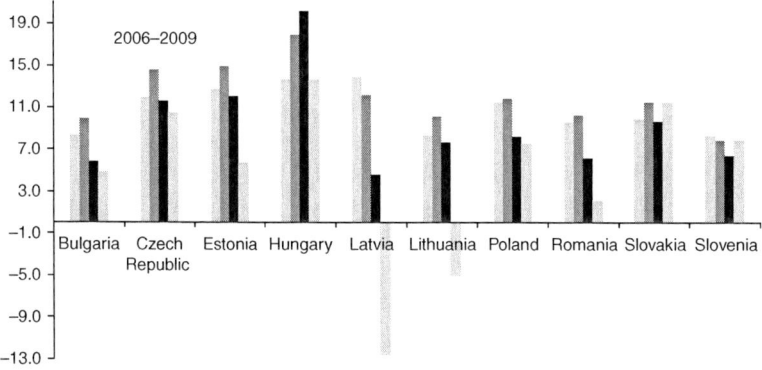

Figure 1.8 FDI-related income outflow (income and its repatriation), in % of inward FDI stock
Source: wiiw, 2010, p. 13.

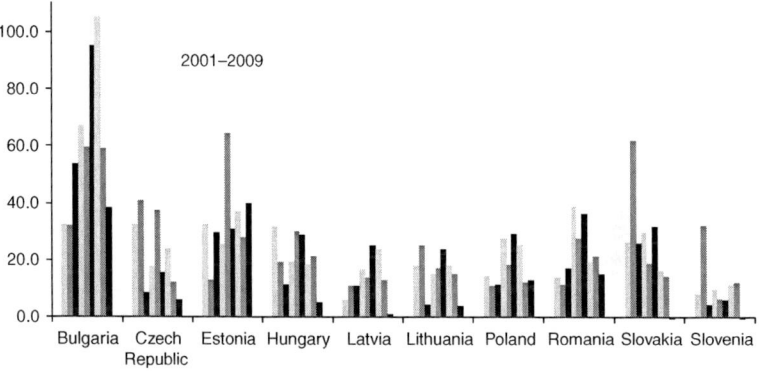

Figure 1.9 Share of FDI inflows in gross fixed capital formation, in %
Source: wiiw, 2010, p. 32.

In the World Investment Report, we have seen that FDI inflows have become a very important source of investment in CEECs that might have even remained dominant for the CEE-10 group had it not been for the capital inflow setback of the recent crisis (see above). The individual CEE-10 members are, however, very heterogeneous: whilst foreign investment is dominant in Bulgaria and very high in Estonia, its share is much lower in the other two Baltics and Slovenia. This suggests that the differences in states of development (or 'depth') of the local currency financial and capital markets may play a part in this result.

It comes as no surprise that most of the FDI inflows into CEE originate from Europe. The USA occupies only sixth place, with a mere 3.9 per cent of all FDI inflows into the region. It is, however, surprising that the Netherlands, Germany, Austria, and France have between them taken up a share of more than 50 per cent of all inflows, especially as this list of the largest European investors includes countries that are geographically rather small, such as the Netherlands and Austria. However, the importance of the latter country for FDI in CEE is explicable due to the common border, the proximity of the capital cities of Vienna and Bratislava and Budapest, and the common history in the dual monarchy with historical ties in industry.

The sectoral distribution of inward FDI into CEECs is a mirror of the structure and development of privatisation programmes. Over all 10 CEECs, inward FDI have accumulated the largest share in manufacturing (30.7 per cent) which is not surprising given the industrial history of the region and its comparatively low wages. This is followed by FDI into financial services (19.2 per cent) which reflects the externally driven development of the banking and direct capital markets. In socialist times, business-related services had been incorporated into the large state-owned firms or conglomerates, whereas other services (other than banking and finance) such as education, health, communication, infrastructure, etc. were at best provided by the state but did not constitute a market for investors. Hence, since transition, investors

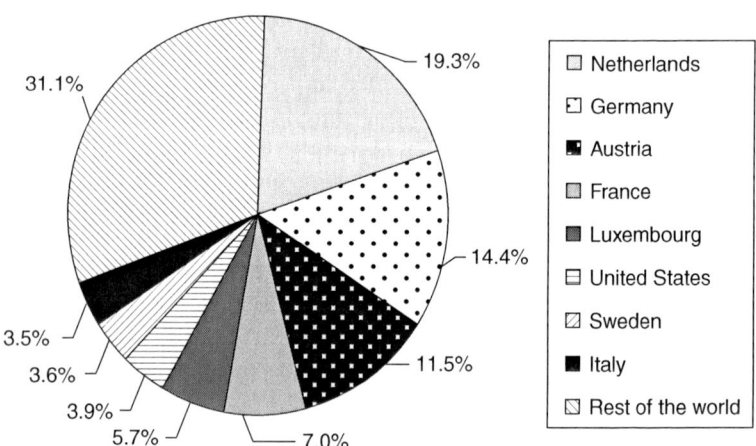

Figure 1.10   Inward FDI stock by major home countries in end 2008, in %
Source: wiiw, 2010, p. 41.

have flocked into services, either buying privatising service parts of state-owned conglomerates or assuming the supply of services where the state has withdrawn. We would therefore expect a dominant share of FDI to flow into services.

FDI outflows can be interpreted as a sign that the firms in the country are sufficiently competitive and profitable to internationalise by investing in a foreign country (the usual theories in international business all assume some 'ownership advantage' as a necessary criterion for foreign investors to bridge the additional costs of setting up an affiliate

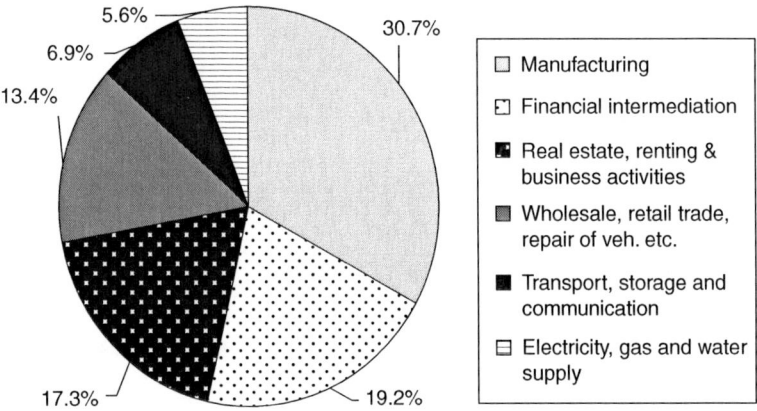

Figure 1.11 Inward FDI stock by economic activity in end 2008, in %
Source: wiiw, 2010, p. 43.

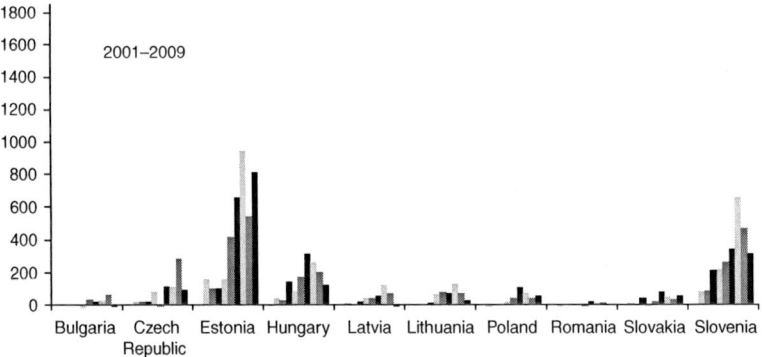

Figure 1.12 FDI outflow, in 1000 EUR mn per capita
Source: wiiw, 2010, p. 26 and EBRD Transition Reports.

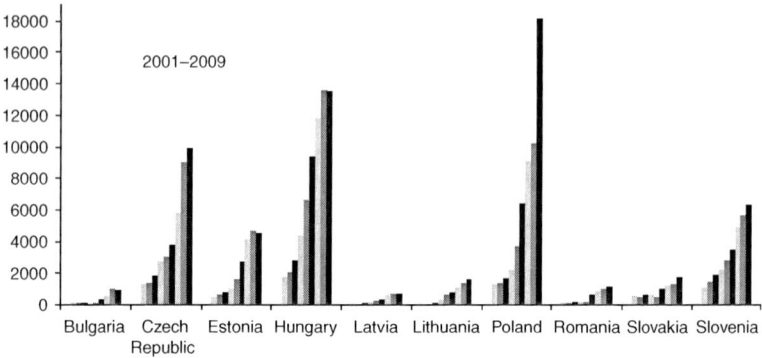

Figure 1.13   Outward FDI stock, in EUR mn
Source: wiiw, 2010, p. 28.

in a foreign country). In the particular case of CEE, this may be valid, but should be discounted by the observation that outward investment often reflects FDI affiliates in the country serving as a springboard for investments 'even further East'.

FDI outflows from CEECs are still much lower than inflows, which is an expected pattern due to the current states of development of those countries. However outflows are already astonishingly high in Estonia (foreign-owned banks use the country as a springboard to further East and other Baltics), Slovenia (the country started with outward, but then allowed only inward FDI, and most outward FDI went to other former Yugoslavian countries, see for example Svetličič, 2007), and Hungary (one thing that is known from the media is that some FDI in Hungary moved on to Romania, so some of the Hungarian outward investment should correspond to Romanian inward investment). The same is probably true for outward FDI from Poland into Ukraine and Belarus.

1.5   Structure of the book

The book consists of nine chapters. Following this introduction, the second chapter describes in some detail the database used in the empirical analysis throughout the other chapters in this book. Chapters 3 through 7 generate the main body of empirical analysis pertaining to the technological aspect of the developmental role of inward FDI. The first of those, Chapter 3 is concerned with the most basic questions in this respect: what are the strategic motives of inward FDI into CEECs, how did they develop, and do foreign investors find a match between their motives and locational conditions? This assumes that a technological role will be more pronounced where strategic motives target the local knowledge base in CEECs and where a match is found between motives, of whatever form, and locational conditions.

Chapter 4 reviews the issue of internal technology transfer, its channels and determinants, and also reviews the empirical results with respect to internal technology transfer in inward FDI into CEECs. This analysis culminates in the development of a taxonomy that will serve as a simple and yet insightful empirical version of the concept of 'developmental subsidiary' in the international business literature. The taxonomy is able to inform about the potentials for static one-way internal technology transfer and the potentials for a dynamic process of reciprocal technological transfer between foreign investor and foreign affiliate. The implications for the technological role of FDI is straightforward: the more dynamic internal technology transfer becomes, the more pronounced will be the technological role.

Chapter 5 and 6 are concerned with external technology transfer and spillovers. Chapter 5 assesses the role of the host innovation systems in CEECs for technological activity of foreign affiliates. This assumes that the technological role of FDI increases with the intensity of local host knowledge sourcing. A large list of potential determinants are identified in the literature, and a subset of those is tested empirically for significance. The resulting set of significant determinants of successful local knowledge sourcing by foreign affiliates is not only informative in itself but can also inform policy-makers about what determines the attractiveness of national innovation systems in their countries for inward FDI beyond the usual tax holidays, provision of infrastructure, and the like. This chapter contains an *excursus* in which the same analysis is reproduced for the case of East Germany – here, however, including spatial determinants characterising the local host environment. This kind of extension is not possible for the CEECs, because of lack of data and data confidentiality with respect to the firm-level database. Chapter 6 reverses the direction of knowledge and technology flows to analyse spillovers from foreign affiliates to the local host economy. Again, the relevant literature is reviewed in order to determine the important channels and determinants of technology spillovers. A subset of determinants is again tested empirically to inform about the significant factors driving spillovers.

Chapter 7 includes the formal institution that is probably the most important with respect to technology transfer and in particular spillovers: the intellectual property rights (IPR) regime. The analysis tries to find out whether a lower intensity of enforcement of IPR rules (as is significantly the case in CEECs) may effect a lessening of the technological role of inward FDI for the host economies by either putting a break on internal technology transfer, or by reducing their own technological activity within the foreign affiliate, or by inhibiting the local host economy cooperation that is so important for the technological role of inward FDI.

Chapter 8 concludes; not in terms of summaries of results generated (those are attached to every individual research chapter), but in terms of an amalgamation of individual results into a broader and more comprehensive picture of what determines the technological aspect of the developmental role of inward FDI. Further, the concluding chapter offers a very brief discussion of possible future avenues to further develop the research programme analysing the developmental, technological role of FDI. This not least provides a view on the caveats and shortcomings in the various sets of analysis and what the missing research foci imply for the future research programme.

As a whole, this work constitutes an attempt to contribute to the identification of the necessary conditions of catch-up development based on technological development driven by inward FDI.

2
The Database Used in the Empirical Analysis

Traditionally, research on FDI analyses bilateral country-level aggregate data on FDI flows (in much the same way as in the introduction). However, the empirical analysis of the technological role of FDI for economic development in host countries (technology transfer and spill-over effects) usually uses the method of production function analysis, based on aggregate industry-level data on FDI stocks combined with estimates of inter-sectoral linkages derived from national-level input–output tables. Some production function analyses rely on firm-specific data using official databases or field work. The traditional method of analysis does not take firm heterogeneity into account, and hence remains unable to explain large fractions of the variance in resulting technological roles between the various FDI projects that exist in host countries. The latter is, however, typically restricted in its comparative strength, because the methodology behind the firm-specific databases varies between the different countries where such data is available. Moreover, official statistics seldom contain information about the most important determinants of the research issues at hand (determinants of the technological role), because they are typically not easily measured in hard facts, but rather involve subjective valuations of managers of firms to be considered in the analysis. Where fieldwork data that includes such non-financial data is used, this often includes only one host country. Most of what the relevant international business literature and international strategic management literature has recently put forward in theoretical or conceptual propositions and empirical insights from tested hypotheses, however, is not traceable by use of official data.

It is Damijan et al. (2003b, 2008) who can be considered to have produced the most comprehensive firm-level studies on FDI produc-tivity spillovers in CEECs; they used data from balance sheets or financial statements, as well as ownership information, from about 91,000 firms in 10 transition economies from 1995 to 2005, sourced from the Amadeus database generated by the Bureau van Dijk, as a tool to evaluate the creditworthiness of firms. Whilst the data provides a large amount of important information on the financial status of a firm and on the kind of ownership of the firm, it does not contain any information about the intensity of interaction of the firms with other firms, or about the management relationship between a foreign investor and its affiliates (the corporate governance issue), apart from the ownership issue that can be traced back as far as the 'ultimate owner'.

2.1   The development of the IWH FDI Micro Database

These shortcomings in the availability of data have motivated the generation of a new database at the firm level, including a selection of CEECs. In a three-year framework, the EU co-financed a research project in the Fifth Framework Programme on the "Determinants of the productivity gap between EU and CEECs" (PRODGAP)[1] involving scientists in the West from the UK, the USA and Germany, and in CEE from Estonia, Poland, the Czech and Slovak Republics, Hungary and Slovenia; this task was initiated in the form of a pilot study involving a concise, two-page questionnaire that reflected some of the most important discussions in the international business literature on technology transfer and spillover determinants. Most of the data generated was non-financial and non-quantitative, but rather qualitative inasmuch as it reflected the subjective valuations of firm managers of foreign affiliates in the region. The topical focus of the questionnaire was 'technological upgrading of foreign affiliates' and generated cross-sectional data from 434 foreign affiliates of manufacturing industries (NACE Rev.1: 15–37) in the five named countries.

In a follow-up project, again co-financed by the EU in its Sixth Framework Programme, on "Understanding the Relationship between Knowledge and Competitiveness in the Enlarging European Union" (U-KNOW),[2] the method and instruments of field work for the database were further refined, and constituted the first wave of the IWH FDI Micro Database[3] (2006–2007). It includes data from 809 foreign affiliates in manufacturing industries of the host economies/regions of East Germany, Romania, Croatia, Poland and Slovenia. The questionnaire was already much larger, again trying to generate only a small amount of financial data, giving way to subjective qualitative data from the perspective of foreign affiliate managers. In topical terms, it was focussed on technology transfer and spillovers/diffusion.

In the same international collaborative project, a second wave was added to the database in 2008, this time involving only East German foreign affiliates. This database is not used in the analyses presented here, because of the country-comparative nature of this book. A third wave, involving East Germany, Romania, Slovakia, the Czech Republic and Hungary was generated in 2008–2009, again in the framework of the U-KNOW project. This wave has generated data for 651 foreign affiliates in the named countries in CEE and another 654 foreign affiliates in East Germany, in manufacturing industries but this time also in selected services. These services include electricity, gas, steam and hot water supply (NACE Rev.1: 40–45), wholesale (NACE Rev.1: 51), transport and financial services (NACE Rev.1: 60–67), computer, R&D, and other business-related services (NACE Rev.1: 72–74), as well as sewage and waste disposal, and media (NACE Rev.1: 90–93).

In 2010, the development of the database was unlinked from its conditionality of availability of external funding via projects and today it forms an integral part of the research base of the IWH. A two-year rhythm for waves involving CEECs is now planned, plus special waves every alternate year focussing on East Germany alone, largely by use of the same questionnaire.

Every wave of the database generated data that can only be used in cross-sectional analysis, because the populations of the samples differ due to a high frequency of entries and exits and the usual difficulties in attracting a high rate of return.

The data in the IWH FDI Micro Database is generated in field studies and hence can be expected to contain some degree of variety purely due to the differing personal perceptions of the interviewees and also to country-specific perceptions (which also may include situations where the interviewer had an influence on the response behaviour of the interviewee). Despite these shortcomings, the database is firmly grounded in international business and strategic management theories and concepts, and constitutes highly useful and qualitative micro data for research into the internationalisation of CEECs and East Germany. As a firm-level database, it is able to account for a large variety of firm, country, network etc. heterogeneity.[4]

2.2 The basic population for the IWH FDI Micro Database[5]

The population for East Germany is drawn from the MARKUS database provided by Verband der Vereine Creditreform e.V.[6] The information in the database is drawn from public indexes, balance sheets, annual reports, the daily press and surveys. It contains about 1.1 million German enterprises; according to Verband der Vereine Creditreform e.V., 97 per cent of all commercially registered and economically active German companies are listed in it. For Germany, these figures seem to be reliable, since any commercial entity is obliged to register with its local chamber of commerce. The database contains enterprise-level information such as name, legal form, date of registration, sector, address, ownership, balance sheet and financial information. It forms the foundation for the population, underlying other established micro datasets such as the Mannheimer Innovation Panel (see Harhoff and Licht, 1993) or the 'KFW and ZEW Start-up Panel' (Fryges et al., 2010) which are both operated by the Centre of European Economic Research (ZEW).

For CEECs, the firm population is drawn from the AMADEUS database provided by the Bureau van Dijk. In total, AMADEUS contains data on 14 million European enterprises and covers 10 transition economies. This data is fully compatible with the information drawn from the MARKUS database; in fact, the latter forms the basis (in a slightly reduced form) for the German part of the AMADEUS database. Bureau van Dijk (2010) describes its AMADEUS dataset as robust against a coverage bias since '35 expert and local information providers assure' the quality of the data (ibid.). Given the compatibility of the MARKUS and AMADEUS databases, it is possible to draw on the population underlying the IWH FDI Micro Database using the following uniform selection criteria for inward and outward FDI in all countries.

The population of enterprises with one or more foreign investors is defined as all enterprises belonging to the selected sectors and countries in the year preceding the field work (2006 for the 2006–2007 wave of the database, and 2008 for the 2008–2009 wave of the database), in which at least one foreign investor holds either a minimum of 10 per cent direct shares/voting rights or a minimum of 25 per cent indirect shares/ voting rights. These enterprises are independent affiliates with their own legal status or they are branches without a legal entity but with their own commercial register entry. Shareholders or ultimate owners are not limited to foreign enterprises headquartered abroad but also include natural persons, donors, foundations, and financial investors with headquarters outside their respective country.

In the case of East Germany, the basic population of enterprises with foreign participation has been supplemented by enterprises belonging to the selected sectors and countries in the year preceding the field work, in which at least one West German multinational investor holds either a minimum of 10 per cent direct shares or voting rights or a minimum of 25 per cent indirect shares or voting rights. A West German multinational investor is defined as an entity that is headquartered in West Germany and has either a minimum of 10 per cent direct shares/ voting rights or at least 25 per cent indirect shares/voting rights in one or more entities located abroad. The federal state of Berlin is considered a part of East Germany in line with other established micro datasets and official statistics (even if that remains an unresolved matter in science).

2.3 Survey sampling and implementation

The sample stratification for the survey in East Germany was proportionally differentiated for ownership (real foreign non-German FDI; West German investment) and further by differentiating between producing industries (NACE REV.1: 10 to 37) and all other industries (NACE Rev.1: 40–45; 51; 60–67; 72–74; 90–93). Subsequently, each of the two sectors was further stratified according to enterprise size in terms of number of employees (up to 9, 10–49, 50–249, more than 250). The sample stratification for the survey in the CEE countries based on the AMADEUS data was broken down by host country and further by differentiating between producing industries (NACE REV.1: 10 to 37) and all other industries (NACE Rev.1: 40–45; 51; 60–67; 72–74; 90–93). Subsequently, those two sectors were further broken down by enterprise size in terms of number of employees (up to 9, 10–49, 50–249, more than 250).

In the 2006–2007 wave, the questionnaire was centrally designed at the IWH and the field work was decentrally organised by following partners: Zentrum für Sozialforschung Halle (zsh) in East Germany, the Institute for Economic Research in Slovenia, the University of Zagreb in Croatia, the Group of Applied Economists in Romania and EMAR Marketing Research in Poland. The sample of firms to be included in the field work was taken out of all firms from the basic population (but in the case of Romania, the size of the population was too large, and a random sample was drawn from the basic population). They were contacted in writing or by phone. The firms were offered a choice between receiving the questionnaire by post or fax, or as an electronic version via email. Most interviews were conducted over the phone, and only a few firms preferred to fill in the questionnaire in written form. The project deliberately allowed the country teams to choose the most appropriate method and timing individually.

The 2008–2009 survey was carried out in a centralised way: the contact addresses and the sample stratification were transferred to the Institute for Applied Social Sciences (infas) for the field work in CEECs and the

Table 2.1 Country composition of the IWH FDI Micro Database, 2006–2007 wave (full) and 2008–2009 wave (CEECs and inward FDI only)

	2006–2007 wave		2008–2009 wave	
	No. of firms	Share of country (%)	No. of firms	Share of country (%)
Croatia	144	17.8	–	–
Czech Republic	–	–	185	14.6
East Germany	295	36.5	654	51.4
Hungary	–	–	57	4.5
Poland	110	13.6	216	17.0
Romania	220	27.2	128	10.1
Slovakia	–	–	30	2.4
Slovenia	40	4.9	–	–
Total	809	100.0	1270	100.0

Source: IWH FDI Micro Database 2006–2007.

Zentrum für Sozialforschung Halle (zsh) for the East German field work. The survey was implemented by means of computer assisted telephone interviews (CATI), which was chosen as the appropriate method because the survey of the IWH FDI Micro Database requires a special design for highly standardised surveys, involves complex target groups, and has substantial filtering in the questionnaire. CATIs are fast and relatively inexpensive, and generate high response rates. In order to further increase the response rate, the enterprise received information about the institute coordinating the field work (IWH), the IWH FDI Micro Database itself, the survey and data confidentiality, by fax and/or email in advance upon request. The questionnaire was first programmed and internally tested for coherency before being submitted to at least five pre-tests per country. The pre-test necessitated minor changes and resulted in a questionnaire which required 15 minutes on average for completion. The interviewers at both providers received intensive training by researchers from the IWH. The interviews were conducted only by native speakers from each country under observation. The interviews were completed in line with the respective sample stratification.

In the 2006–2007 wave, the total population for East Germany and the CEE countries included 1412 and 5421 enterprises respectively. For East Germany, a total sample of 295 interviews could be conducted, corresponding to a response rate of 20.9 per cent. In the case of CEE countries, 514 interviews could be carried out, corresponding to a response rate of 9.5 per cent. Thus, a total of 809 enterprises participated in the 2006–2007 survey for the IWH FDI Micro Database. This generates an overall response rate of 11.8 per cent.

In the 2008–2009 wave, 2815 East German and 6801 CEE companies were contacted during the survey. For East Germany, a total sample of 654 interviews could be conducted, corresponding to a response rate of 23.1 per cent. In the case of CEE countries, 651 interviews were carried out, corresponding to a response rate of 9.6 per cent. Thus, a total of 1270 enterprises participated in the 2008–2009 survey for the IWH FDI Micro Database. This generates an overall response rate of 13.6 per cent. The analysis in this book only uses the CEE part of this wave, not the East German sample, and only uses the FDI outward part of the database. This corresponds to 616 foreign affiliates in the Czech Republic, Hungary, Poland, Romania and Slovakia to be included in the analysis.

2.4 Survey representativeness of the two IWH FDI Micro Database waves

2.4.1 The 2006–2007 wave of the IWH FDI Micro Database

From the total population of firms, about 12 per cent could not be contacted by interviewers by telephone; in most cases this was related to an incorrect telephone number (see IWH, 2007, Table A6 for a complete list). About 67 per cent of firms refused to participate in the survey and can be classed as non-respondents. Non-response was motivated by explanations such as no interest, time constraints, refusal to be interviewed by telephone, and postponement of a possible interview to a later stage. Finally, about 21 per cent of firms agreed to participate in the survey (this includes the interviews during the pre-test). The latter group of firms can be classified as respondents. The tests for significant differences in the distribution of firms across sectors, regions, and firm size categories were done comparing respondents to non-respondents, giving an indication of the extent to which the survey suffers from a non-response bias.

In the East German sample, Chi-square tests (see IWH, 2007) show that the sample is representative at the sectoral level[7] but differs significantly from the total population with regard to regional[8] and size distribution. The regional deviations are mainly related to an under-representation of firms from Berlin, and firms with 10 to 249 employees are over-represented. Moreover, there are some, even though weak, indications for a non-response bias. An additional limitation applies, because representativeness was tested by looking at each criterion (sector, region, size) separately and not jointly. Beyond these criteria, the sample is characterised by affiliates that entered throughout the period between 1990 and 2005. It is dominated by multinational enterprise groups as owners, full ownership as well as acquisition being the mode of entry.

In the CEE sample, deviations of the sample from the distribution of the population across sectors reach maximum levels of 3 per cent, counted by the number of firms. In terms of employment by sector, deviations turn out to be up to 5 per cent. The sample response rates vary across countries (see IWH, 2007). In terms of number of firms, the range is from 6.6 per cent in Poland to 65.5 per cent in Croatia. In terms of employment, the range is from 11.1 per cent of the sample of West German multinational-owned firms in East Germany to 66.0 per cent in Croatia. Such differences in response rates are mainly explained by the difference in the size of the respective total population, the larger populations of Romania, Poland, and East Germany tending to show lower response rates. Yet the share of each country in the resulting sample does reflect the respective size of populations across countries. In terms of distribution of firms across size classes (in terms of employees), the sample is under-represented for micro (1–9) and small enterprise (10–49), and consequently over-represented for medium-sized (50–249) and large (above 250) firms.

2.4.2   The 2008–2009 wave of the IWH FDI Micro database

About 28 per cent of East German enterprises and 14 per cent of CEE enterprises could not be contacted due to reasons such as wrong contact numbers, insolvency or incorrect other information (see IWH, 2009, Annex Table 1 for a complete list).

For the sample of multinational investors in East Germany, the representativeness analysis shows that the distribution does not differ significantly from the underlying population with regard to sectors (producing industries and all other industries) and ownership structure (full, majority or minority multinational-owned) (see IWH, 2009). In contrast, we find significant differences for the regional distribution (at the level of the federal states as well as at the regionally even more disaggregated levels of the so-called *Raumordnungsregionen*) of industries (NACE two-digit level), and firm size (up to 9, 10–49, 50–249, more than 250 employees). These significant differences are caused by the sample of foreign-owned firms in East Germany, while the sample of West German-owned firms is representative with respect to all dimensions tested. Among East German enterprises with a foreign investor, the regional sample deviation is mostly driven by the strong under-representation of enterprises located in Berlin. It is worth while pointing out that the regional distribution was not part of the sample stratification. Furthermore, there is an under-representation in the sample of companies with more than 250 employees.

As for the FDI sample in CEECs, the tests show significant differences in the distribution across the five countries due to under-representation of Czech and Polish firms and the corresponding over-representation of Hungarian, Slovakian and Romanian firms (see IWH, 2009, Table 3). For each individual country sample, the analysis finds no significant deviation in the sectoral distribution (producing industries and all other industries) between population and sample. This also applies to the industry distribution (at NACE two-digit level). The only exception here is the Hungarian sample. In addition, no significant differences in the firm size distribution between population and sample for the transition economies are detected. In general, the results suggest that the population and its corresponding samples generate a reliable micro database. The survey is representative of various indicators; therefore, it meets the relevant criteria for scientific research within this field.

2.5 Survey questionnaires

The theory of technological accumulation and firm internationalisation (Cantwell, 1989) proposes a dynamic relationship between spatially bounded technological externalities, and the internationalisation of firms' R&D and innovation, as well as the potential for technological spillovers from MNEs to the domestic economy. This type of theorising was crucial for the design of the IWH FDI Micro Database, which looks at the role of MNEs in selected transition economies as well as East Germany from a comparative perspective.

In the 2006–2007 wave, the thematic focus of the survey was placed upon investment motives and the evaluation of locational factors. The questionnaire includes 23 questions[9] and is divided into four sections. The first part of the questionnaire (questions 1–8) covers standard information about the foreign investor, type of investor, headquarters location, date of entry, mode of entry and investment motives. The second part (questions 9–13) focuses on the foreign affiliate and includes questions on employment, turnover, sales and procurement structures and intermediate inputs, as well as performance indicators and how those have developed in the preceding three years. The third part (questions 14–17) deals with the interaction and relationship between the foreign affiliate and its investor, such as autonomy over particular business functions, and changes in mandate that have occurred and that are expected to occur in the future. The last part of the questionnaire (questions 18–23) deals with the technological activity of the foreign affiliate (R&D, innovations, innovation effects on sales), sources of technological activity, sources of knowledge (own R&D, actors in the host innovation system, external innovation systems, and the foreign investor's own network) for technological activity within the foreign affiliate, and finally an evaluation of importance of the foreign affiliate itself as a source of knowledge for technological activity amongst actors in the host innovation system, external innovation systems, and the foreign investor's own network. All R&D indicators are in line with the international standards as codified in the Frascati Manual (OECD, 2002); all innovation indicators are in line with the Oslo Manual (OECD, 2005).

In the 2008–2009 wave, the thematic focus of the survey was investment motives and the evaluation of locational factors. The questionnaire includes 38 questions[10] and is divided into five sections. The first section (questions 1–5) mainly covers the evaluation of locational factors. These are broken down into traditional factors such as quantitative labour supply and the availability of government grants and subsidies, as well as the potential for technological cooperation. In addition, 'soft' locational factors, including culture, image, health care, and availability of housing, have been evaluated by the participating firms. The first part of the questionnaire proper (questions 6–12) also includes standard questions about the structure of the foreign investor with questions on the type of investor, location of headquarters, date of entry, mode of entry and investment motive, as well as autonomy over particular business functions. The second part (questions 13–20) is answered by enterprises with outward FDI only. The third part (questions 21–26) deals with questions about research and development (R&D), including changes to R&D employment through internationalisation and R&D cooperation. All R&D indicators are again in line with the international standards as codified in the Frascati Manual (OECD, 2002). Part four of the questionnaire (questions 27–30) deals with product innovations, including their intensity, and changes to product innovation intensity through the internationalisation processes. All innovation-related indicators are again in line with the international standards as codified in the Oslo Manual (OECD, 2005). The final part of the questionnaire includes questions on employment, turnover, intermediate inputs, exports as, well as changes to selected performance indicators through the internationalisation processes.

Both questionnaires are reproduced in the annexe.

3
Foreign Direct Investment Motives and the Match with Locational Conditions in Central East Europe

The analysis in this chapter focuses on the two central aspects of inward FDI: strategic investment motives and locational factors. Their relative importance over host countries in CEE, regions of origin, year of engagement etc. is described to present a more realistic picture of the dominant investment motives prevailing in CEECs. Next, investment motives are compared to locational advantages to test the extent of correspondence at the firm level between what the motives imply in terms of locational factors and the motive itself. This is derived from theory of international production (e.g. Cantwell and Iammarino, 1998, 2003) and the international management literature (e.g. von Zedtwitz and Gassman, 2002; Andersson et al., 2002) that shows that the strategy followed by a foreign investor depends not only on internal considerations but also on the endowment of the host economy or region with locational factors.

If FDI were to assume a significant role in the technological catching up of a host country, then this would depend on whether foreign investors find that locational conditions correspond to their strategic investment motives. Whereas a rational investor can be expected to have evaluated locational conditions before entering a market by investing in equity, conditions may on a closer look turn out to be different than expected, and conditions may also change over time. In particular the latter is important in CEECs with their dynamic changes to every aspect of economic activity that started with the opening towards the West and still continues, even if at a lower rate of change. It is therefore not obvious whether FDI projects that have generated sunk costs in CEECs today feel that their strategic motives are still (or have by now as expected) matched their investment motives.

Of course, the technological role of FDI in the host economy depends on the kind of strategic motive followed by FDI projects, some being more prone and others less so to technology transfer and diffusion. And yet any kind of strategic motive-led project can have positive effects on technology transfer and diffusion, and even the motives most prone to technology transfer and diffusion may result in no such effects or even negative ones. This is particularly likely where motives are not matched by locational conditions. Hence a satisfactory analysis of strategic motives has to extend to asking whether or not hopes and expectations tied to the strategic motive are in the event fulfilled.

Therefore the two main research questions to be asked here are:

RQ(3_1) What do/did foreign investors expect from their investments in the region?

RQ(3_2) Do/did investors find a match between their strategic motives and the locational advantages in their host economies?

To answer these questions, this chapter first summarises what the relevant literature presents in terms of dominant investment motives in CEE and on the determinants of motives. Then the strategic motives of foreign investors investing in CEE is described by use of the IWH FDI Micro Database. Since there can be considerable differences in the motive to invest between types of home countries, the analysis distinguishes between three groups of countries where the investment originated: European countries, developed countries outside Europe, and emerging markets. The complete sample is then used to test for changes in investment motives over time (1989 to 2009). During that period, the CEECs transformed from centrally planned economies with rather low levels of economic development to fully fledged market economies integrated in the European common market and with high growth rates. The evolution of investment motives over time should reflect that transition process. Next, investment motives are broken down according to the location of foreign affiliates in the individual CEECs. This indicates where differences lie and whether they correspond to the general impression of these countries. Finally, the analysis tests the match between the quality of locational advantages in the host economies with the strategic motive of the investor.

3.1   Current state of the art in research

In terms of theory, the choice of location by multinational firms is influenced by a range of different factors, which are partly external and partly internal to firms. Dunning's OLI concept (Dunning 1977, 1981, 1993, and 2009) may provide insights where it is concerned with the interaction of mainly ownership and location advantages: a firm will in its localisation decisions attempt to align the advantages that particular localisations offer with its firm-specific set of ownership advantages (or may assume that it is in a position to contribute to changing location advantages to better complement its ownership advantages: Cantwell's interaction between locational and ownership advantages: Cantwell, 1989, and 2000; Cantwell and Piscitello, 2005). The assumed relationship between localisation and ownership advantages then gives rise to the investment motive that a particular firm has established for itself when investing in a particular location.

More recently, agglomeration theories have become more important when trying to explain locational choices by TNCs (see e.g. Fujita and Krugman and Vanables, 1999). Their main points of interest include the existence and shape of localised externalities and the effects of competition between foreign and domestic firms. The literature on locational preferences of foreign investment additionally considers that locational choice depends not only on the types of activities of firms or the characteristics of the host country, but also on the multinational firms' investment motives (e.g. Dunning, 2009; Enderwick, 2005). Dunning's approach to what attracts FDI focuses on four primary investment motives that provide a widely used classification on investment motives: resource-seeking, market-seeking, efficiency-seeking, and strategic asset-seeking investment. The efficiency-seeking motive can be further differentiated into cost advantages regarding factors of production, economies of scale and economies of scope (Dunning and Lundan, 2008).

There is ample empirical research on strategic investment motives in CEECs. Amongst the many contributions, the following may be representative of the knowledge available on the issue (of course there are many more, but it is the objective of this brief survey to provide a general overview with particular reference to the research question of this chapter): Andreff and Andreff (2005) use panel data of official macro/meso statistics between 1993 and 2003 to establish that inward FDI in CEECs is driven by market size and purchasing power, and also by lower unit labour costs and other determinants that include institutional variables such as the banking system, the economic significance of the state and the quality of governance. At the firm level, Lankes and Venables (1996) use survey data for 145 FDI projects by 117 multinational companies in CEECs, and find that national and regional market access is the predominant investment motive. Manea and Pearce (2004) also use survey data to analyse motives for investment in transition economies, and distinguish between market seeking, efficiency seeking, and knowledge seeking as potential motives for FDI. Among the 135 multinational affiliates surveyed, more than 78 per cent rated market seeking as the primary motive for investment. This result is supported by cross-country surveys such as Meyer (1998), Svetličič and Rojec (1994) and Altomonte (2000), as well as survey studies of FDI in Slovenia (Rojec and Svetličič, 1993), and Italian investors in CEECs (Mutinelli and Piscitello, 1997). Even in a fairly recent survey, Paul and Wooster (2008) find that US investors in CEECs tend to follow market-seeking objectives.

Contemporary research, however, increasingly suggests that dominant investment strategies are difficult to single out for any host country or region by considering all investment projects as one homogeneous group: rather, heterogeneities in terms of country of origin of the foreign investor, the sector, the year and mode of entry of the foreign investor etc. have an important influence on investment motives.

Chung and Alcacer (2002) suggest that the perceived weight that foreign investors attach to the different strategic investment motives influence their assessment of quality of locational factors. Mudambi and Mudambi (2002) assess the TNC's decision between greenfield vs acquisition and find that not only do firm-specific factors play a role (such as their own R&D intensity or the investor's preference for or against diversification or concentration), but that investors likewise consider locational factors such as industry growth and market power in the host country and industry.

Other analyses of locational factors concentrate on either specific host countries or a specific home country of the foreign investor. Basile (2005), who also assesses greenfield vs acquisition, finds in the case of Italy that the main location determinants differ markedly according to the type of foreign investment considered. Disdier and Mayer (2004) analyse location choices by French firms in Eastern and Western Europe; their analysis focuses on a possible East–West divide and finds that location decisions in this respect are strongly influenced by the institutional quality of the host country. Boudier-Bensebaa (2005) investigates FDI in Hungary at a regional level, while Chidlow et al. (2009) do so for FDI in Poland. Some contributions select single determinants for choice of location, such as access to technology or the influence of research and development (R&D) activity (Chung and Alcácer, 2002) – or, focusing on the CEE countries, labour costs (Bellak, et al. 2008), or the institutional framework in the host country (for example Bevan et al. 2004; Meyer and

Jensen 2004). Other surveys investigate the determinants of FDI in CEECs at the sectoral level showing that sector-specific factors affect the choice of the final location (Resmini, 2000; Pusterla and Resmini, 2007). Using an example of Italian investors in selected CEE countries, Majocchi and Strange (2007) find that not only are market size and growth, the endowment with labour, and infrastructure important determinants of FDI, but also financial and market liberalisation. For the following analysis, therefore, it is important to consider such potential heterogeneities.

3.2 The data used for the analysis

The CEE-section of the IWH FDI Micro Database of 2008–2009 contains information from 185 foreign affiliates in the Czech Republic, 57 in Hungary, 216 in Poland, 128 in Romania, and 30 in Slovenia. The previous waves of the database do contain strategic investment motives, but no information about locational conditions. This was an important focus of the 2008–2009 wave. This data allows the distinction to be made between six distinct strategic motives for investment: (i) market access; (ii) cost advantages of production factors (which mainly relate to the comparatively low wage levels in the region as compared with Western Europe); (iii) economies of scale; (iv) product diversification in the foreign investor's network; (v) access to local knowledge and technology; and (vi) access to localised natural resources. The data is ranked on an ordinal Likert scale ranging from 1 for "not important at all" to 4 for "very important" in increments of 1. Firms were asked to provide an indication of their opinion about the level of importance for the foreign investor on each of the six motives and at the time of investment into their firm. To give the data a time dimension, the database additionally contains information about the year of investment.

In terms of locational factors, the IWH FDI Micro Database considers 14 locational factors, which are grouped in four classes:

1. Quantitative supply of labour (low-qualification workers, apprentices (trainees), junior employees with university degrees, qualified employees)
2. Availability of state support (investment subsidies, financial incentives for R&D and innovation)
3. Potential for technological cooperation (with local public and private science institutions, with other local firms)
4. Socio-cultural environment (culture on offer, supply of health services, supply of housing and accommodation, no hostility towards foreign

workers, supply of child-minding facilities, image of the region in general).

This data is ranked on the same Likert scale as the data above on strategic investment motives, except that here it ranges from "very bad" (1) to "very good" (4).

The database contains not only manufacturing industries, but also selected service industries (see the complete list in Chapter 2.1 The development of the IWH FDI Micro Database).

3.3   Strategic motives of foreign investors investing in Central East Europe

Over the whole population of investors in CEECs and disregarding possible heterogeneities, the strategic motive of reaping cost advantages from production factors and the motive of market access dominated the assessment of importance by foreign investors in the CEECs at the time of entry of the foreign investor (see Table 3.1). This largely corresponds to the general picture in the literature on CEECs. With regard to the costs related to labour, all CEECs still offer sizeable advantages today that are not completely eroded by equally low levels of labour productivity.[1]

Of slightly less importance are the efficiency motives of scale economies and product diversification, whereas the motive of access to local knowledge and technology rates surprisingly high considering that the countries were largely isolated from Western technological developments during their socialist era. Access to localised natural resources play a very limited role for investments into the CEECs (developments over time are considered in Table 3.2).

Explicitly considering heterogeneity in terms of host countries results in far fewer differences than anticipated.[2] The results are astonishing not least because home countries were grouped according to probably the most important sources of heterogeneity, that is their level of economic development and their proximity to the EU. Still, some differences are identified: as can be expected, investors from emerging markets place less emphasis on cost advantages and scale economies because they can easily enjoy those advantages at home. Rather, the motive of product diversification appears to be more important (and significantly so). Investors from emerging markets appear to use their locations in the CEECs to improve their product mix, possibly complementing the established, standardised products that they produce at home in larger quantities.

Table 3.1 Investment motives of foreign investments into CEE, distinguished by region of origin

	Total population	West European countries [a]	Emerging markets [b]	Non-European developed c. [c]
	$n=604$	$n=506$	$n=47$	$n=51$
	mean	deviation	deviation	deviation
Market access	3.11	0.04	-0.04	-0.04
Cost advantages	3.14	0.01	-0.10	0.08
Scale economies	2.91	0.09	-0.16	-0.00
Product diversification	2.70	-0.17	**0.47***	-0.13
Access to technology	2.72	-0.17	0.05	**0.25***
Access to natural resources	1.99	-0.07	0.12	0.02

Notes: The mean is defined as the simple average of the investment motive. Deviations are defined as differences between the mean of the specific country group category and the mean of the rest of the population. n is the number of firms that have provided information. Coefficients with a * and in bold are significant at the .1 level.
a) Austria, Belgium, Denmark, Finland, France, Germany, Greece, Iceland, Ireland, Italy, Liechtenstein, Luxembourg, Netherlands, Norway, Portugal, Spain, Sweden, Switzerland, United Kingdom.
b) Belize, Croatia, Cyprus, Czech Republic, North Korea, Hungary, Lebanon, Lithuania, Malta, Netherlands Antilles, Poland, Romania, Russian Federation, San Marino, Slovak Republic, Slovenia, South Africa, Turkey, Ukraine.
c) Canada, Israel, Japan, United States.
Source: IWH FDI Micro database 2008–2009.

3.3.1 Investment motives by regions of origin

Other notable differences emerge from the analysis, despite not being statistically significant: in contrast to emerging markets as home countries, the sign for product diversification appears negative for the European and for other, non-European, developed economies. More importantly, the strategic motive of access to technology also appears less important for investors from West European countries, whereas it is relatively more important (and significant) for the group of non-European developed countries. Investors from non-European developed countries follow different objectives when investing in the CEECs than do their West European competitors; Western European firms tend to rely on nearby home technology bases, whereas non-European investors see more

Table 3.2 Investment motives, distinguished by time of entry of the foreign investor, 1989–2009

	1989–1995	1996–2000	2001–2006	2007–2009
	n=162	n=186	n=160	n=108
	mean	deviation	deviation	deviation
Market access	3.31	−0.22*	−0.11	0.04
Cost advantages	3.14	0.06	−0.04	−0.21*
Scale economies	2.87	0.04	0.06	−0.08
Product diversification	2.57	0.10	0.09	0.09
Access to technology	2.71	−0.06	**0.20***	−0.21*
Access to natural resources	1.96	0.01	0.08	−0.03

Notes: The mean is defined as the simple average of the investment motive. Deviations are defined as differences between the mean of the category during the period of observation and that of the previous period. n is the number of firms that have provided information. Coefficients with a * and in bold are significant at the .1 level.

Source: IWH FDI Micro database 2008–2009.

opportunities to make profitable use of the local or national innovation systems in CEECs. In other words, space may explain local technology sourcing, but adaptation of the product to the local market conditions and environment in the host economy will play a more important role in this context (see Boudier-Bensebaa, 2005, on differences in regional attractiveness of FDI in the case of Hungary).

These results suggest that FDI may play an important role for technological development in the CEECs, where foreign affiliates seek economically useful technical knowledge in their host location. This appears to be the case for FDI originating from non-European developed countries, even if technological levels can be expected to be lower in CEECs: it is the essence of TNCs to shop around their locations for differences in all factors or production, including technologies (consumer tastes, climatic conditions, production methods). CEECs are hence probably not as detached from Western technological developments as traditionally believed and, over time, appear to have been able to generate their own technical knowledge specific to the region and valuable even to developed countries. The observation that this is not the case for FDI originating from West European countries furthermore suggests that CEECs may be an interesting location to get access to West European knowledge and technology, even if owned by West European investors in CEECs (which eventually raises the question of protection of intellectual property rights, see Chapter 7). The foreign affiliates' positive trend in absorptive capacities (the issue of absorptive capacities is discussed in detail in Chapter 4 on the conditions of internal technology transfer) may explain much of this trend, but more importantly, CEECs are increasingly able to make commercially valuable use of their long and uninterrupted industrial history and the technology they were able to get access to precisely because of their successes in attracting FDI in the more recent past.

3.3.2 The development of investment motives over time

Of course, investment motives will have changed over time – and possibly most importantly in line with the growth in labour costs (or the decline in real effective labour unit cost advantages over locations in the West). In the earlier wave of the database (2006–2007), it was also the motive of market access that turned out to be clearly dominant by a large margin with a mean of 3.42. This was followed by the motive "to increase efficiency across the foreign owner network" (which includes labour unit cost advantages) with a mean of 3.16, and "access to local technology", already ranking third with a mean of 2.91. Because market access is of such high importance in the region, and because local demand rises with wages and purchasing power, not much change between the different motives can be observed over time. The 2008–2009 wave of the IWH FDI Micro Database, however, implicitly contains data on the development of strategic motives from the outset of systemic transition until recently: Table 3.2 illustrates the evolution of strategic motives for investment in the CEECs. Four periods are distinguished, covering the transition from a centrally planned economy within the Soviet sphere to a transformed competitive economy integrated into the European market:[3] first, the period of systemic transition that began in 1989; the second period, starting in 1996 when most of the large-scale privatisation had been completed in most transition countries; the third period, beginning in 2001 when economic development already relied on a broader base of a mix of large and small and medium-sized firms and FDI was already less focused on privatisation projects; and the period of integration and financial crisis, beginning in 2007. In the process of systemic transition and economic development, one would expect the strategic motives to change, in particular from cost- and efficiency-related advantages to more localised technology advantages. Alas, this will take a long time, and the period of observation here may be too short to observe this yet.

The results indicate that the cost-advantage motive did not change much until the mid-2000s but was significantly less important in the final period.[4] However, there was no corresponding increase in importance of the technology-related motive. Rather, it appears that the technology-related motive became relatively even less important in the third period. The loss of this advantage may however be due to the current financial crisis, where investors tend to reallocate the more sensitive business functions into the headquarters to reduce short-term risks. But this may be temporary, as transnational corporations often postpone R&D projects, or shift certain functions within the enterprise, until the global economy recovers. In general, however, they do tend to develop their R&D strategy with a long time-horizon. Since the survey only covers the initial stages of the financial crisis, it is too early to tell whether the trend will be long-lived.

An important question that arises from the results is whether the trends above in general, and the adverse trend away from technology sourcing in the CEECs, can be explained by changes in the geographic origin of the investment. As suggested in Table 3.1, non-European investors were comparatively more technology-seeking, whereas European investors put less emphasis on this motive. Alas, it emerges from additional analysis of the data (not reported here) that the share of non-European investment in the CEECs increased from 8.6 per cent in period three to 12.0 per cent in period four; likewise the share of investments from emerging markets rose from 5.0 per cent to 11.0 per cent between those two same periods. While the recent decline in the technology-related motive may be the result of foreign investors centralising more risky technological activities nearer to their headquarters, it remains an open issue as to whether the quality of the national innovation system in the host economies has declined, or whether a change in industrial structure of foreign investments lies at the heart of this result.

3.3.3   Investment motives by host economies

Investment motives are not homogeneous across different host economies. Host countries with particularly low (unit) wage costs, such as Romania, tend to attract investors who place more weight on the motive of tapping cost advantages, whereas large host countries, such as Poland or other countries with sufficient purchasing power on the domestic market, such as the Czech Republic, should expect to attract interest in market-seeking FDI. Countries with better developed innovation systems in general and public and private science sectors in particular (mainly the Czech Republic, but also possibly Poland and Hungary) should be more attractive to those investors who are seeking localised knowledge and technology.

Table 3.3 suggests that foreign investors into Romania consider cost advantages to be the most important investment motive. The strategic motive of tapping localised knowledge and technology is significantly more important in the Czech Republic than in the other host countries. This also applies to Slovakia, although this deviation remains statistically insignificant. Interestingly, foreign investors in Poland appear to place much less weight on the role of the national innovation system for their own technological development – the average weight is sizeably and significantly lower than in the rest of the population. This does not shed a positive light on the attempts of the Polish policy to transform and develop its actors in the national innovation system. Equally, the strategic motive of market access is significantly less important in Poland than the rest of the population in the other host countries. Whilst this is also a puzzling result given the large size of the country, it is perhaps the lower level of purchasing power (due to lower wages in Poland than in the other CEECs, apart from Romania) that drives this result.

Access to natural resources did not play an important role for the population as a whole. Still, much higher levels (with significant deviation) are attached to this motive in Hungary and Romania. In Romania,

Table 3.3 Investment motives, distinguished by host economies

	Czech Republic	Hungary	Poland	Romania	Slovakia
	n=185	n=57	n=216	n=128	n=30
	deviation	deviation	deviation	deviation	deviation
Market access	0.13	−0.01	−0.23*	0.08	0.28
Cost advantages	−0.06	−0.09	−0.10	**0.24***	0.10
Scale economies	0.03	0.10	−0.07	−0.05	0.17
Product diversification	0.13	−0.16	−0.13	0.17	−0.21
Access to technology	**0.29***	0.05	**−0.35***	0.04	0.22
Access to nat. resources	**−0.18***	**0.45***	**−0.19***	**0.28***	−0.10

Notes: Deviations are defined as differences between the mean of the particular country category and the mean of the rest of the population. n is the number of firms that have provided information. Coefficients with a * and in bold are significant at the .1 level.

Source: IWH FDI Micro database 2008–2009.

this may be a result of lignite mining, a resource that is also available elsewhere, but relatively low wages and possibly weak labour protection regulations are important in this country.

3.3.4   Investment motives by industry classification

The distinction between firms engaged in manufacturing production or the provision of services is an important determinant of investment motives. Table 3.4 shows that market access is significantly less important for firms in manufacturing industries, while for the same group of firms cost advantages are significantly more important. The reason may be that firms in service industries have to be on site while manufacturing firms also have the option to access a market by exporting largely prefabricated products. Cost advantages are significantly more important to manufacturing firms, because the quantity and variety in qualification of the personnel needed tends to be higher than that of firms in the services selected in the database. The significantly positive difference in importance of economies of scale to manufacturing firms might be due to the fact that these firms are more likely to adapt their products to the new market. The difference in importance of product diversification is slightly negative when compared to firms in manufacturing industries, but not statistically significant. Firms in service

Table 3.4   Investment motives, distinguished by industry classifications

	Selected services (A)	Manufacturing (B)	(B) – (A)
	n=279	n=337	
	mean	mean	deviation
Market access	3.34	2.92	−0.42*
Cost advantages	2.93	3.30	0.38*
Scale economies	2.79	3.00	0.21*
Product diversification	2.74	2.67	−0.07
Access to technology	2.77	2.68	−0.09
Access to natural resources	1.92	2.04	0.12

Notes: The mean is defined as the simple average of the investment motive. Deviations are defined as differences between the mean of the manufacturing category and the mean of the category selected services. n is the number of firms that have provided information. Coefficients with a * and in bold are significant at the .1 level.

Source: IWH FDI Micro database 2008–2009.

industries tend to adapt more intensively to the characteristics of the new market. Access to technology, too, is slightly less important for manufacturing firms, while access to natural resources tends, naturally, to be more important.

3.3.5 Endowment with locational advantages

The analysis of the quality of locational factors is conducted both for the entire population and for each of the four CEECs individually. Table 3.5 shows that some of the factors in the socio-cultural environment assume the highest quality levels in the assessment of FDI across all the CEECs. This is followed by the potential for technological cooperation and the quantitative supply of labour. The locational factor of availability of state support is comparatively the weakest. Foreign investors, indeed any rational investor, will usually tend to accept state support. The important question, however, is whether investors are attracted because of the subsidies or whether they take an interest in the host economy that goes beyond financial subsidies. Not least, European state aid rules strictly limit the possibilities in this respect. Distinguishing between the five host economies, the data reveals that foreign investors in Romania are the least satisfied with the provision of state support, for both categories of investment subsidies and financial incentives for R&D and innovation. This may indicate either a lower quality in the provision of state support (that is more or less granted for all investments into the EU), or that investors in Romania expected even more support to counterbalance other locational disadvantages. In contrast, investors in Hungary are significantly more content with the quality of state support – the deviations are statistically significant for both categories.

Satisfaction with the supply of both workers with low qualification levels and of qualified personnel appears to be fairly high (in particular, higher than for apprentices and for trainees and junior employees with university degrees) with little variation between host countries. In all four categories of employees, foreign investors in the Czech Republic are less content – and significantly so with the supply of apprentices, trainees and university graduates. In Hungary and Poland, however, university graduates turn out to be significantly better supplied than the rest of the total population. In Poland this applies equally to trainees, and in Romania young university graduates appear to be more difficult to acquire.

When comparing manufacturing industries to services, the main disparities arise with regard to the 'soft' locational factors (see Table 3.6); manufacturing firms are significantly less satisfied with the provision of cultural offers, supply of health services, supply of housing and

Table 3.5 Endowment with locational advantages, distinguished by host economy

	Total population	Czech Republic	Hungary	Poland	Romania	Slovakia
	n=616	n=185	n=57	n=216	n=128	n=30
	mean	deviation	deviation	deviation	deviation	deviation
Quantitative supply of labour						
Low-qualification workers	2.98	−0.00	0.08	−0.01	−0.01	−0.07
Apprentices (trainees)	2.67	−0.17*	−0.03	0.14*	−0.00	0.10
Junior employees with university degrees	2.53	−0.17*	0.29*	0.15*	−0.17*	0.08
Qualified employees	2.94	−0.04	−0.07	−0.06	0.12	0.14
Availability of state support						
Investment subsidies	2.31	0.11*	0.37*	0.10	−0.43*	0.01
Financial incentives for R&D and innovation	2.33	0.11	0.35*	0.24*	−0.58*	−0.05
Potential for technological cooperation						
Local public and private science institutions	2.80	0.22*	0.16	0.03	−0.36*	−0.03
Other local firms	3.05	0.05	0.10	−0.03	−0.12*	0.17*

Socio–cultural environment

Culture on offer	3.03	0.13*	0.13	−0.02	−0.28*	0.29*
Supply of health services	2.81	0.39*	0.09	−0.27*	−0.22*	0.13
Supply of housing and accommodation	3.03	−0.12*	0.16	0.07	−0.07	0.15
No hostility against foreign workers	3.42	−0.01*	0.23*	−0.03	−0.02	0.25*
Supply of child-minding facilities	2.83	0.15*	0.07	0.13*	−0.05	0.02
Image of the region in general	3.13	0.07	−0.19*	0.04	−0.05	0.00

Notes: The mean is defined as the simple average of the locational advantages. Deviations are defined as difference between the average of the country-specific group of firms and the average of all other firms. n is the number of firms that provided information. Coefficients with a * and in bold are significant at the .1 level.
Source: IWH FDI Micro database 2008–2009.

Table 3.6 Endowment with locational advantages, distinguished by industry classification

	Total population	Selected services (A)	Manufacturing (B)	(B) – (A)
	n=616	n=279	n=337	
	mean	mean	mean	deviation
Quantitative supply of labour				
Low-qualification workers	2.98	2.93	3.03	**0.10***
Apprentices (trainees)	2.67	2.72	2.63	−0.09
Junior employees with university degrees	2.53	2.58	2.49	−0.09
Qualified employees	2.94	3.03	2.87	**−0.16***
Availability of state support				
Investment subsidies	2.31	2.31	2.31	0.00
Financial incentives for R&D and innovation	2.33	2.36	2.31	−0.05
Potential for technological cooperation				
Local public and private science institutions	2.80	2.89	2.72	**−0.17***
Other local firms	3.05	3.06	3.04	−0.02
Socio-cultural environment				
Culture on offer	3.03	3.18	2.91	**−0.27***
Supply of health services	2.81	2.88	2.76	**−0.12***
Supply of housing and accommodation	3.03	3.11	2.96	**−0.15***
No hostility against foreign workers	3.42	3.41	3.43	0.02
Supply of child-minding facilities	2.83	2.82	2.84	0.02
Image of the region in general	3.13	3.19	3.09	**0.10***

Notes: The mean is defined as the simple average of the locational advantages. Deviations are defined as difference between the average of the manufacturing category and the average of the Selected Services category. n is the number of firms that have provided information. Coefficients with a * and in bold are significant at the .1 level.
Source: IWH FDI Micro database 2008–2009..

accommodation, and the general image of the region chosen by the foreign investor. The reason may be that manufacturing firms often locate in the periphery of agglomerations (commercial parks), where socio-cultural offers are less pronounced, whereas service firms often locate in the core of agglomerations (towns and cities) with a more lively socio-cultural environment. Manufacturing firms also appear to be less satisfied with the endowment in all categories of labour force except for that of low-qualified workers, where satisfaction is significantly higher than that of foreign affiliates in service industries. Demand for qualified personnel is higher in the manufacturing sector while there is a greater need for lower qualified workers in services (offering often standardised services designed in the home countries of the foreign investor). Manufacturing firms are also significantly less satisfied with the potential for technological cooperation with local public and private science institutions.

3.3.6 The match between locational advantages and investment motives

The comparison of the relative importance of investment motives with the quality of locational factors (the question of a match) does not distinguish between host countries or other possible sources of heterogeneity, because this would make reading and interpreting results very difficult.[5] The comparison between investment motives and locational advantages is based on plausibility assumptions as to what investment motives can be assumed to be asking for which locational advantages. This provides a straightforward way to assess the match between strategic motives and fulfilment of expectations of foreign investors or investments in terms of perceived quality of locational factors. For example: do firms that rate cost advantages as very important also find the local quantitative supply of labour in the different employee categories to be of a sufficient quality? Or: do technology- and knowledge-seeking foreign investors find fruitful potential for technological cooperation with actors in the innovation systems of their host economies/regions?

Table 3.7 asks whether the foreign affiliates' evaluations of qualities of locational advantages in the host economies correspond to their individual strategic investment motives (the matching question). In fact, the foreign affiliates that hold that cost advantages are "very important" also assess the supply of low-qualified personnel as being significantly better than do all the investors for which cost advantages are "important", "less important", or "not important". This, however, is not the case for the employment categories of apprentices and of junior employees

Table 3.7 Endowment with locational advantages, distinguished by strategic motives

	Market access	Cost advantages	Scale economies	Product diversification	Access to technology	Access to natural resources
	n=254	n=225	n=172	n=161	n=130	n=78
	deviation	deviation	deviation	deviation	deviation	deviation
Quantitative supply of labour						
Low-qualification workers	0.04	0.14*	0.01	−0.10	0.02	−0.03
Apprentices (trainees)	0.09	−0.07	0.04	0.03	0.12*	0.07
Junior employees with university degrees	0.03	−0.06	0.03	−0.06	−0.06	0.00
Qualified employees	0.26*	0.10	0.08	0.18*	0.12	0.03
Availability of state support						
Investment subsidies	0.02	0.01	−0.02	0.01	0.07	0.03
Financial incentives for R&D and innovation	0.02	0.06	0.13*	0.05	0.15*	−0.01
Potential for technological cooperation						
Local public and private science institutions	0.03	0.01	0.05	−0.03	0.10	−0.02
Other local firms	0.14*	0.04	0.09*	0.09*	0.03	0.14*

Socio-cultural environment

Culture on offer	**0.16***	−0.01	0.10	0.08	**0.27***	−0.03
Supply of health services	0.09	−0.06	0.02	**0.21***	**0.14***	0.06
Supply of housing and accommodation	**0.11***	0.04	0.09	**0.19***	0.05	0.03
No hostility against foreign workers	**0.18***	**0.09***	**0.18***	0.04	0.03	0.04
Supply of child-minding facilities	0.08	−0.03	**0.11***	**0.14***	**0.16***	0.06
Image of the region in general	**0.13***	0.03	0.09	**0.15***	**0.19***	0.02

Notes: Deviations are defined as difference between the average of the group of firms who value the strategic motive as "very important" and the firms that value this motive as either "important", "less important", or "not important". n is the number of firms that have provided information. Coefficients with a * and in bold are significant at the .1 level.
Source: IWH FDI Micro database 2008–2009.

with university degrees (with negative deviations to 'the rest', even if not significant). Nor is this the case for qualified employees (with a positive sign, yet not significant). The negative deviations for the two middle-qualification employee categories suggest a typical mismatch: the expectation of foreign investors is not fulfilled. Cost-advantage seekers also find state support to be of better quality than do all other foreign investors – this, however, with only a small margin and not significant.

Cost advantages were particularly relevant to investors in Romania from the point of view of the host country (Table 3.3). State support in Romania was felt to be particularly weak, and the assessment of the supply of labour is mostly negative (Table 3.5) – potential signs of mismatches. Foreign investors into Slovakia, also comparatively more cost advantage-seeking, appear to be quite happy with the supply of labour (apart from labour at the lower end of the qualification spectrum), even if not with significant deviations.

Table 3.8 probes directly into the matching question, and the analysis is at the firm level: a positive (partial[6]) correlation between the firms' individual levels of importance of strategic motives and the evaluation of locational conditions shows a positive match, negative results did not occur. Whilst negative correlations would have been entirely possible, such a result would have represented a situation in which the manager of the foreign affiliate follows a strategic motive that is diametrically opposed to the locational conditions. In the cases of low-qualification

Table 3.8   Partial correlations between strategic motives and locational conditions

Strategic motive cost advantage ...	obs.	Partial correlation	significance
Low-qualification workers	616	**0.15**	**0.000**
Apprentices (trainees)	616	0.06	0.112
Junior employees with university degrees	616	**0.14**	**0.000**
Qualified employees	616	**0.10**	**0.011**
Investment subsidies	616	**0.10**	**0.012**
Financial incentives for R&D and innovation	616	0.06	0.150
Strategic motive access to technology ...			
Local public and private science institutions	616	**0.10**	**0.015**
Other local firms	616	0.02	0.648

Note: The results for the other determinants included in the partial correlation analysis are presented in the Appendix to Chapter 7, Table A5a.

workers, junior employees and qualified employees, as well as for investment subsidies as a form of state support, the correlations are in fact statistically significant, even if the correlation coefficients are rather low (which is often the case in partial correlation analyses with heterogeneous firm-level data).

Foreign investors seeking access to localised technology and knowledge are expected to find a good match with their strategy if they positively assess the quality of potentials for technological cooperation with actors in the host economies. Whilst this appears to be the case generally (positive deviations, more so with regard to scientific institutions, not so much with other firms), the deviations are, however, not statistically significant. It seems that local market-seekers, foreign investors trying to reap economies of scale, as well as those looking for access to natural resources, rate technological cooperation potentials higher than, in fact, do technology-seekers. Foreign affiliates in the Czech Republic were particularly interested in localised technology and knowledge, and were also happier with the potential for cooperation with local science institutions (statistically significant). This suggests the kind of positive match that policy can take as a strong momentum in their efforts to attract FDI. Other local firms were also, yet only mildly, important for the fulfilling of the expectations of foreign investors. In Slovakia, there also appears to be a positive match, yet with the opposite ranking between the two potential cooperation partners: foreign affiliates also look out for access to localised knowledge and technology, yet seem to find the potential with scientific institutions less interesting than with other local firms (and significantly so). Further research is necessary to shed more light on these results.

Finally, foreign investments that search for local technology, focus on product diversification and are market-seekers place particular weight on those soft locational advantages that often prove to be so important in the decision making of foreign investors.

The (partial) correlation analysis at the firm level shows that it was in particular cooperation with local public and private scientific institutions that produced a positive and significant match between this strategic investment motive and locational conditions.

3.4 Summary of main results

The analysis of investment motives largely confirms the results in other literature that market access and cost advantages play a dominant role for FDI projects in CEECs. Three reasons are noticeable in driving this pattern: first, the economies of CEE are in the process of integration with the EU; countries that are not yet (or were not at the time of investment) EU members do have free trade agreements (for example in the institutional set-up of the CEFTA) and hence offer access to the European markets for non-European investors. Second, at the outset of the transition process, the economies of CEE had very little internationally competitive supply, and demand was expected to catch up swiftly to Western levels, making their markets interesting targets for investment aiming at the opportunities of those supply gaps. Third, wages are much lower than in the West, and even if productivity also exhibits large gaps, unit labour costs still offer some advantage. And yet, the countries are endowed with well trained and educated industrial employees. Rising wages do not appear to significantly affect the market-seeking strategic motive over the 20-year period, whilst the cost-related motive has in fact become less important.

This is not matched by a corresponding increase in importance of the motive to gain access to localised technology, a motive that would exceed a positive technological role of FDI in their host economies. The importance of this motive has actually dropped significantly between 2007 and 2009. Whether this is an effect rooted in the global financial crisis, or whether this signals important deficiencies in the innovation systems of the hosting countries remains an open issue.

In a comparison between the different host countries in CEE, access to localised technology is much less important in Poland, and more important in the Czech Republic and Slovakia. In Poland, market access is of less importance than in the other CEECs, which suggests that market seeking is probably more related to the purchasing power of potential buyers than to the sheer size of the market in terms of number of buyers. This adds a dynamic perspective to the assessment of managers in the region, which may be rooted in the domestic growth rates that typically are above the rates in the West, and have been particularly high in the case of Poland.

The analysis also suggests that investment motives are not specific to the host country and its economic environment alone, but rather that further determinants, heterogeneities, have to be considered to provide a more robust picture. In this respect, foreign affiliates in manufacturing industries are more interested in cost- and scale-related motives, whereas affiliates in service industries focus more on market access. Moreover, investors from emerging markets are more interested in using FDI to diversify their product portfolios, and non-European developed countries see more opportunities to make profitable use of the innovation systems in CEECs.

This analysis does provide some insightful answers to research question RQ(3_1) on what foreign investors expect from their investments in the region. With regard to RQ(3_2) on whether investors find a match between their strategic motives and the locational advantages, the analysis produces interesting mixed results: satisfaction amongst foreign affiliates with the endowment with locational factors is high for the quantitative supply of labour across all CEECs. This is particularly true for Poland's supply with employees at medium levels of qualification, whereas in the Czech Republic, this is clearly less so, and Hungary appears to have a particularly good supply of young university graduates. Those results do correspond to the respective countries' differences in dominant strategic motives. With regard to the availability of state support, foreign affiliates in Romania are less satisfied (which corresponds to the dominance of cost-related motives), whilst investors in Hungary are more satisfied. Plausibly, foreign affiliates in manufacturing industries are more satisfied with the lower stratum of employee qualification than are foreign affiliates in service industries, but less so for higher levels of qualification. Manufacturing foreign affiliates are also less satisfied with the potential for technological cooperation with local public and private science institutions than foreign affiliates in service industries.

The analysis finds a positive match between investors seeking cost advantages and the quantitative supply of low-qualification workers and qualified employees, whereas there may be a mismatch for trainees and junior employees with university degrees, that is the middle ground in terms of qualification of employees. In the case of Romania, this potential mismatch becomes particularly obvious. Moreover, the analysis produces positive, even if weak, signs of a match between local technology-seekers and potentials for local technological cooperation, which is particularly strong in the case of the Czech Republic (with scientific institutions) and Slovakia (with other domestic firms).

Finally, the results point to the apparently counterintuitive situation: it is not the technology-seeking foreign affiliates that value highly the potentials for technological cooperation with local institutions and firms, but rather the market-seekers and FDI projects looking for economies of scale. This either suggests some mismatch or else may signal negative experience with the technological capability of the host economy: expectations of technology-seeking investments were not fulfilled, whereas market-seekers without such expectations may have found it useful to cooperate with the local economy. Again, further research is needed to generate a more robust interpretation.

4
Conditions of Internal Technology Transfer and Spillovers between Foreign Investors and Foreign Affiliates in Central East Europe

Whereas the previous chapter was concerned with motives and matches between expectations and reality by foreign investments in CEECs as a condition of technology transfer and spillovers, Chapters 4 to 6 separately analyse the conditions for internal (Chapter 4) and external (Chapters 5 and 6) technology transfer and spillovers. Amongst the most widely discussed determinants of the technological role of FDI for host economies is internal, direct technology transfer within the network of the foreign investor (including the intentional contracted, and the unintentional, not accounted for in contracts, that is spillovers).

The part of the international business literature using the original 'pipeline' model (at least implicitly – this concept goes back to Marin and Bell, 2006) assumes the headquarters to be the originator of knowledge and technology and sees foreign affiliates at the receiving end (for example, Caves, 1974, and Globerman, 1979, using industry level and cross-sectional research designs; Haskel et al., 2002, and Keller and Yeaple, 2003, with firm-level studies using panel data). Whilst this view easily lends itself to theoretical modelling, empirical analysis has increasingly opened the realm of possibilities to internal transfer without the participation of headquarters. Further, this literature also assumes the transfer of knowledge and technology to go in either direction between the foreign affiliate, other firms in the foreign investor's network, and headquarters.

In light of the empirical objectives of this chapter, the analysis considers both directions of transfer of knowledge and technology.

It does not grant the foreign investors' headquarters a unique role in the technology diffusion processes, nor is any distinction made between headquarters and other firms in the foreign investor's network.

The basic assumption is that if FDI were to assume an important role in technological advance of host economies, then this would not simply be based on the trickling down of knowledge and technology from the foreign investor to its affiliate. Rather, this role is, importantly, propelled by a reverse direction of learning, whereby the foreign investor networks benefits from access to the knowledge and technology of the foreign affiliate.

The main research questions of this chapter are hence:

RQ(4_1) What are the dominant determinants of the intense internal technology transfer that goes in both directions?

RQ(4_2) What are the most important of such determinants for the case of CEECs that suffice to produce an insightful representation of the countries' potentials to benefit from internal technology and knowledge transfer via its inward FDI projects?

RQ(4_3) Given the conditions prevailing in CEE as countries hosting FDI, what are the relative potentials for an intense internal technology transfer that goes in both directions for each individual country?

The analysis in this chapter aims to contribute to answering those questions and proceeds by summarising what the relevant literature holds on the channels and determinants of internal technology transfer, and what it's empirical fraction holds on the issue of internal transfer. Then, the two-dimensional conceptual taxonomy is developed and finally filled with data on foreign affiliates in CEE to assess the potentials of each CEE host country to benefit from dynamic, two-way internal technology transfer via inward FDI.

4.1 The current state of the art in research

4.1.1 The channels of internal technology transfer and spillovers

Amongst all possible channels of international technology transfer (international trade in goods and services, international trade of patents and licenses, international mobility of labour, demonstration effects by way of imitation and reverse engineering, joint ventures, etc.), FDI probably forms the most important one (see for example Blomström and Kokko, 1998) and the cheapest one (see the discussion of transaction costs, see also Damijan et al., 2003b). It may also be the safest channel, because it allows the protection of intellectual properties by exerting control based on ownership, rather than arm-length control by reliance on the intellectual property regime (IPR). FDI may also transfer more recent technology and more quickly than would licensing agreements or even international trade (Mansfield and Romeo, 1980). FDI can transfer knowledge and technology embodied in products and services (superior machinery, patents), as well as in the form of intangible assets such as organisational and management skills (methods for quality management, scientific and engineering techniques, or market research, new procedures, requiring integrative learning and coordination, etc.) (see for example Child, 1993; Kinoshita, 2000; Meyer, 2003; Kvinge, 2004).

A further important channel of internal technology transfer involves the training of personnel of the host country affiliate at all possible levels (see for example Blomström and Kokko, 2003; Meyer, 2003). Other capabilities, such as the skill to access and select the most relevant external technology not only depend on the foreign affiliate, but also on the environment and the level and intensity of interaction between the foreign affiliate and its environment (Meyer, 2003). Interestingly, foreign investors often tend not to transfer the part of knowledge that would be necessary to generate own independent innovations or to make own strategic decisions (Meyer and Jensen, 2004). This will be particularly relevant for host countries with lower levels of protection of IPRs, which is also the case in CEE (see Chapter 7 specifically on this issues).

4.1.2   The determinants of internal technology transfer

The question of what determines the amount of technology transferred by FDI projects is discussed mainly in the international business and management strategy literatures. At the most general level, internal technology and knowledge transfer depend on the characteristics of knowledge, of the foreign investor and its affiliate(s), and on the characteristics of the home and host countries.

First and foremost, and with the oldest roots, the 'technology gap hypothesis' holds that the potential for technology transfer (both internal and external) increases with the size of the technological gap between the home country of the foreign investor and the host country of its affiliate. This idea goes back to Veblen (1915) and in particular to Gerschenkron's 'advantages of backwardness' (1952): backward countries with lower levels of technology are assumed to more easily find more advanced technology than countries already at the technological frontier. The distinction between innovation and imitation may be an illustrative case in point here. This view had been further qualified by for example Perez (1997), who holds that to benefit from such an advantage of relative technological backwardness, foreign penetration has to be modest and slowly growing, and that technological catching up can be further propelled by industrial policies in support of R&D. Subsequently, the technology gap hypothesis has been subject to further criticism (for example Cantwell, 1989), suggesting that large gaps may make foreign technologies too different from the ones in the host economy, including the foreign affiliate. Up until today, there is no universally agreed opinion or empirical indication as to the exact relationship between the technology gap and the potentials for internal (and also external) technology transfer. There is both empirical evidence in support of the linear hypothesis (for example Castellani and Zanfei, 2003, for selected EU-15 countries, Blomström and Wolff, 1994 for Mexico, and Blalock and Gertler, 2004, for Indonesian firms) as well as empirical evidence at variance with the linear version of the hypothesis (for example Kokko, 1994, for Mexico; Kokko et al., 1996, for Uruguay; Imbriani and Reganti, 1997, for the Italian manufacturing sector).

Following on from this, the literature uses the concept of 'absorptive capacity' to qualify the above hypothesis, focussing on the human capital necessary to make valuable use of the technology that a firm has gained access to (for a comprehensive review of the literature on absorptive capacities, see for example Clark et al., 2011).[1] The theoretical concept is often attached to Cohen and Levinthal (1989, 1990) which is based on the idea of learning, in which R&D not only stimulates own technological development (for example possibly leading to innovations), but also increases a firm's capacity to absorb external knowledge and technology. Those two faces of R&D help to develop the ability of firms to recognise and identify valuable new knowledge and technology, to assimilate and integrate it into the firm, and to thereby exploit and use it productively in the own firm and thereby to increase the firm's own knowledge base. Focussing on the conditions of learning within an organisation (interorganisational learning), Lane and Lubatkin (1998) suggest that the determinants of learning between firms (similarity of the partners' basic knowledge, lower management formalisation, research centralisation, compensation practices, and research communities) is a better measure of absorptive capacity than is R&D spending. It goes without saying that the strength of the Cohen and Levinthal concept over that by Lane and Lubatkin is a much simpler operationalisation in empirical research. The trade-off is that the former is unable to describe the conditions of learning and improvements of absorptive capacities. Further related conceptual research includes Keller (1996) and Borensztein et al. (1998), focusing on technology accumulation and human capital in the firms, and Van den Bosch et al. (1999) who assess the firm's organisational structure and combinative capabilities.

A slightly different approach is developed by evolutionary theories and the cognitive approach: the idea of 'organisational learning' focuses on the firm's ability to acquire new knowledge from its environment (for example Levitt and March, 1987; Kogut and Zander, 1996; Marcotte and Niosi 2000; Nahapiet and Ghoshal, 1998). This is related to the issue of absorptive capacities, and distinguishes between different characteristics of knowledge. It also adds a dynamic perspective: where knowledge received is explicit and codifiable, practice through learning-by-doing may suffice to incorporate external knowledge. If, however, knowledge assumes a more tacit character, then the recipient firm has to actively try to interpret correctly what it has received and has to be able to understand the fundamentals of the technology used. Importantly, it is not only the firm at the receiving end that is obliged to learn how to adopt the knowledge received – but also the original owner of tacit knowledge to be transferred likewise has to learn how to adequately teach the part of the knowledge that they are typically not aware of as being relevant, just because of its tacit nature. In the context of Polanyi (1957): they know more than they are *able* to tell (p. 44).

In empirical studies, absorptive capacities are often measured traditionally by R&D intensity in Cohen and Levinthal-style (for example Kinoshita, 2000; Kneller, 2002; Barrios et al., 2003), or by proxies such as investment into intangible assets (Damijan and Knell, 2003), or human capital endowment and training (Schoors and van d. Tool, 2002; Todo and Miyamato, 2002; Kneller, 2002). Blalock and Gertler (2004) proxy absorptive capacities by R&D and also control for human capital endowment as well as the productivity gap between domestic and foreign firms. The basic assumption is that technology transfer grows in intensity with the level of absorptive capacities. Those empirical studies generally establish a positive relationship between technology transfer and absorptive capacities. In particular, Girma and Görg (2005) use UK establishment data to suggest a U-shaped relationship between productivity growth and FDI interacting with absorptive capacity: technology transfer effects on productivity depend on simultaneously high levels of FDI and absorptive capacities, whilst FDI or absorptive capacities alone may not suffice to make a noticeable difference. Interestingly, the U shape also suggests that if the amount of FDI remains moderate, the effects of technology transfer may still be large despite weak absorptive capacities – a result that corresponds to the qualification of the technology gap hypothesis above. In contrast to those positive findings, Castellani and Zanfei (2003) and Barrios and Strobl (2002) find no evidence for a positive relationship between absorptive capacities and spillovers in their respective studies.

A further important determinant of internal technology transfer pertains to the characteristics of knowledge and technology itself, Mansfield and Romeo (1980) find that if the technology is fairly modern and complex, knowledge and technology is transferred within the MNC network to a greater extent than to outsiders (for example competitor, customer or supplier). Kogut and Zander (1993) distinguish between the extent of tacitness of knowledge, and conclude that tacit knowledge is transferred preferably to fully-owned foreign affiliates rather than to joint ventures. Licensing becomes a preferred option where knowledge is easily codifiable and teachable. In their conceptualisation, a firm can be characterised as a 'repository of social knowledge that structures cooperative action' (ibid., p. 627) and TNCs' networks 'are social communities that specialise in the creation and internal transfer of knowledge' (ibid., p. 625, see also 627). The transfer of R&D within the network of the foreign investor to its affiliate[2] is probably the highest form of internal technology transfer. And still it remains somewhat limited in emerging markets and is difficult to identify on a broad scale in CEECs – very little is known about its actual size amongst CEECs as hosts to FDI projects.

Investment motives, such as the ones analysed above, are also obvious determinants of internal technology transfer: amongst the six investment motives considered above, the strategic asset-seeking motive of access to localised knowledge and technology is the one in which internal technology transfer may be assumed to be most intense: tapping local knowledge bases can be assumed to be closely connected to both absorptive capacities and the reciprocity of exchange of knowledge and technology – both of which correspond to internal technology transfer. Where, however, the dominant strategic motive is geared towards accessing the host market, the transfer of knowledge and technology from parent to foreign affiliate will typically be confined to production and service processes (marketing, logistics etc.). Moreover, the impact on productivity within market-seeking projects may be large at the beginning, but will abate swiftly after the distribution channels have been established (see for example von Tunzelmann, 2004, p. 33). Adding to the variety of different kinds of FDI, very little internal technology transfer can be assumed for so-called extended workbenches or outward processing trade (OPT):[3] OPT uses (labour) cost advantages by way of international fragmentation of production to gain a competitive advantage over competitors from low-cost countries (sometimes also called 'screwdriver assembly plants', see for example von Tunzelmann, 1995, p. 371, on the case of East Asia during their early catching-up phases). In the European context, preferential trade arrangements with CEECs for OPT additionally played an important role during the early phases of transition and European integration (see for example Andreff and Andreff, 2001a and 2001b; Boudier-Bensebaà and Brezinski, 2000, Pellegrin, 1999).[4] As such, OPT will not involve much intentional internal technology transfer beyond what is necessary to reap cost advantages by an efficient production using complementary factor intensities. Arguably, however, OPT may still involve some technological benefits for the host economy (see Senior Nello, 2002): it may enable the 'foreign production unit' to benefit from higher quality inputs in the form of prefabricated products (even if the main aim is to process those and subsequently send them back to their origin with little potential for other use). OPT may help to penetrate foreign distribution networks and gain knowledge of working to international standards, quality requirements, and marketing (pp. 298–99). In sum, however, the implication for internal technology transfer from OPT is that rather little technology with a long-term value is intentionally transferred to the assembly plant by its investor.[5] And even where this is still done, the foreign investor tends to make sure that as little knowledge and technology as possible will dissipate to the host economy.

Transfer of R&D and investment motives both give rise to another very important determinant of internal technology transfer, and that is the management relationship between the two partners in the internal technology diffusion process: parent and foreign affiliate. Also the issue of absorptive capacity reviewed above leads directly to this management relationship: by enhancing its adaptive capabilities, the foreign affiliate may establish a process of technological interaction to the benefit of both partners, but it needs some autonomy in deciding about its own business functions to actually be able to do so. Organisational theory and the international business and management strategy literature offer important insights into this determinant (see in particular: White and Poynter, 1984; Young, Hood and Dunlop, 1988; Bartlet and Ghoshal, 1989, 2004; Birkinshaw and Hood, 1998; Tavares, 2001; and Holm, Malmberg and Sölvell, 2002). In these bodies of literature, multinational investors are characterised as differentiated 'inter-organisational networks' (Roth and Morrison, 1992, p. 141) in which a variety of different foreign affiliates often operate as 'quasi firms' (Tavares, 2001). Within such a network, the role of the foreign affiliate determines the mechanism and extent of control. Hence, 'the subsidiary is a semiautonomous entity capable of making its own decisions but [is] constrained in its action by the demand of head office managers' (Birkinshaw and Hood, 1998, p. 780). It is in particular this strand of literature that culminated into the concept of MNC, or subsidiary evolution and subsidiary development (or 'developmental subsidiary'), in which foreign affiliates may gradually mature into centres of excellence (ibid., p. 141–211) and may even assume 'technical leadership' (Frost, 2001) – if all conditions are in support of this, the transfer then becomes reciprocal.[6] In terms of industries, technical leadership is mainly to be found in pharmaceuticals, semiconductors, and the manufacture of photographic instruments, where the dominant share of patents are attributable to the foreign affiliate.[7]

Finally, the formal institutional environment plays an important role as determinant of internal technology transfer. Policies incentivising FDI in general and internal transfer, R&D and innovation in particular, legal regimes protecting physical property and IPRs etc. all contribute to an institutional environment conducive to internal technology transfer. In the case of CEECs, the legal aspect of this environment (the 'rule of law') is in fact well developed and supportive of intense internal technology transfer due to the introduction of the *acquis communautaire* into the legal body of each member state, and the gradual takeover of this in the countries having applied for EU membership.[8] Alas, there is an important difference between having laws in the books as formal institutions on the one hand and their interpretation and use – the degree of enforcement of the law – on the other. Here, even CEECs show important gaps; the IPR regime is used exemplarily in Chapter 7 of this work.

4.1.3 Empirical analysis of internal technology transfer in CEECs

In the body of literature on technology transfer through inward FDI into CEECs, most studies support the assumption that internal technology transfer does in fact happen and is significant in the statistical sense (for a review of the literature, see Jindra, 2006, pp. 36–53). In the case of CEECs, this is often related to privatisation, assuming that foreign ownership provides the necessary efficient corporate governance whilst firms privatised domestically or even to insiders are expected to lack expertise, incentives (see for example Blanchard, 1997; Damijan et al., 2003b), and possibly even capital (the financial and capital markets in CEECs are only slowly becoming capable of providing capital for investment). The typical empirical analyses use firm-level panel data and test positively that foreign affiliates have experienced faster growth in productivity than domestic enterprises since the foreign investor started its engagement. Negative internal technology effects via FDI in CEECs are identified in only one study (Damijan et al., 2003b) and only for the Czech Republic and Poland, not for the other eight CEECs of the study (Bulgaria, Estonia, Hungary, Latvia, Lithuania, Romania, Slovakia, Slovenia).[9]

Not only does the empirical literature on CEECs suggest positive internal technology transfer, but also the results show that internal transfer tends to be much larger than external effects (see Chapter 6). In Damijan et al. (2003b), internal effects are larger by a factor of 50 as compared to backward linkages to upstream suppliers in the host economy, and larger by a factor of 500 as compared to horizontal linkages to host economy firms of the same industry. Other attempts to quantify internal technology transfer effects calculate sales to be higher by some 8–16 per cent amongst firms with foreign presence (Bosco, 2001). Evenett and Voico (2002, p. 10) find in a study of the Czech Republic that total factor productivity amongst firms receiving FDI increased by a sizeable 43 per cent between 1994 and 1998 (a period which followed substantial privatisation, often involving foreign investment). They hold that foreign investors tend to engage in firms of dominant size and such that are above-average successful on the relevant market, that is 'picking winners' (see also Djankov and Hoekman, 1998). In contrast, Damijan et al. (2003b) suggest that foreign investments were targeted more at capital-intensive and human skills-intensive investments than at high-productivity or large-size ones. In Djankov and Hoekman (1998), the foreign-ownership share in the affiliate turns out to be positively related to internal technology transfer, whereas this is not supported by Schoors and van der Tool (2002) and Damijan et al. (2003a). Damijan et al. (2003b) find a positive effect of majority ownership, apart from the cases of Lithuania and Romania.

With respect to the technology gap hypothesis, Torlak (2004) finds larger internal technology effects in Bulgaria, Poland, and Romania (with lower initial productivity levels) than in the Czech Republic and Hungary. This interpretation, however, does not always hold in other studies, where the more advanced CEECs show larger internal effects.

Reviews of the literature suggest that one of the factors decisive for the quality of the analysis of such tests is the explicit consideration of possible selection biases. In fact, most of the relevant studies do so, apart from

Bosco, 2001, and Torlak, 2004 (which are nevertheless reviewed here, because of their otherwise valuable results). All studies reviewed show clearly that the industrial sector of the investment matters: this is mostly accounted for by way of industry dummies, and in the case of the Evenett and Voico (2002) study, industry effects are explicitly considered in their analysis. The following analysis also duly accounts for the industries of the foreign affiliates.

4.2 The 'role in the network' and absorptive capacity: a two-dimensional concept

The conceptual taxonomy developed in the following has the purpose of providing a framework for the analysis of the development or maturing of foreign affiliates with a view on the potentials for internal technology transfer that goes in both directions. In addition, the taxonomy has to lend itself to an empirical representation by use of firm-specific data contained within the IWH FDI Micro Database. The concept is quite close to the concept of 'developmental subsidiary', even if much less refined (due to the empirical imperative of the analysis). The conceptual taxonomy proposed here may be viewed as one possible empirical application of the concept of 'developmental subsidiary'.

Amongst the many conditions for internal knowledge and technology effects of FDI discussed in the literature (see above), the main focus here is on the absorptive capacities of the foreign affiliate. The central role of this condition for the three questions of the chapter becomes particularly apparent when considering the example of a foreign investor seeking to invest into a firm in a country with a technology gap, as for example from West to East Europe: only if the foreign investor's affiliate has sufficient experience of the technology used by the investor will its affiliate be able to make productive use of the foreign knowledge and technology and thereby contribute to the spurring of productivity growth in the host economy. In the adverse case, an affiliate with little or no absorptive capacity will have an effect solely on employment and income growth (if at all, which obviously depends on the extent of the replacement effect of the foreign investment). Many other conditions are important, certainly, but previous research does suggest that it is mainly absorptive capacity that has proved to be a particular bottleneck in Central and East Europe (Dyker et al., 2006, pp. 79–82; see also Dyker, 2006; Varblane et al., 2007). With respect to the relationship between absorptive capacities and prospects for internal technology transfer, plausibility assumptions suggest that the potential for internal technology transfer will increase with the ability of the affiliate to absorb its foreign investor's knowledge and technology, and to adapt these to its local environment conditions (see Girma and Görg, 2005).[10]

But the analysis presented here aims to reach beyond this focus on absorptive capacities alone and adds the role that the affiliate assumes within the foreign investor's network to assess potentials of internal technology transfer simultaneously with absorptive capacities. The typical assumption is that a foreign affiliate will be able to outcompete domestic enterprises with the assistance of the foreign investor. This, however, not only depends on the ability of the affiliate to absorb and make use of the potential advantage that the foreign investor holds in the form of its particular knowledge and experience (Dunning's ownership advantage), but also on the interorganisational setup of the relationship between foreign investor and affiliate – its autonomy in deciding about vital business functions with a view on the conditions in the foreign affiliate and its environment, a view the foreign investor often has less intimate knowledge of than do the managers on site. This conceptualisation not only adds another account of firm-heterogeneity to the analysis of internal technology transfer, but also combines the two factors that promise to generate particularly interesting new insights into the conditions of intense internal technology transfer.

The 'role in the network' is conceptualised by the management relationship between 'head office managers' and managers of the foreign affiliate: a dominant parent will manage its affiliate on its behalf without much interference from the affiliate's own management. At the other extreme, an autonomous foreign affiliate is characterised by a mandate of managing its own future plans, while the parent takes an inactive management role. A dominant parent will tend to implement its foreign technology in its affiliate, whereas an autonomous affiliate will tend to assume a more active role in this process. The literature on the role of affiliate autonomy on internal technology transfer is not fully congruent: in the early stages of the development of foreign affiliates, parent companies can be 'adverse to technological incongruity' (Dyker and Stolberg, 2003, following Ozawa, 1979, and Wells, 1983) and could 'tend to place considerable stress on the importance of being able to impose their own technological culture on subsidiaries (...) as a way of guaranteeing control over productivity' (Dyker and Stolberg, 2003, p. 4). Kokko and Kravtsova (2008) find in their analysis of innovative capability in foreign affiliates in four transition countries (Estonia, Hungary, Poland, Slovenia) that capabilities tend to increase with the level of independence of the affiliate from its foreign investor. Meanwhile, Damijan et al. (2011) find that the transfer of responsibilities from headquarters to affiliates is conducive to process innovation.

Installation of alien technology without its adaptation to the specific features of the host country environment will, however, make the process stagnate in a static one-way technology transfer. The process ends with the installation of the parent's 'best practice' in its foreign affiliate, regardless of whether the technology functions efficiently in the particular environment of the host economy. It becomes dynamic with the foreign affiliate 'maturing' in terms of its own expertise and gradually assuming a more active role in the adaptation of the parent's technology. In a process of technological interaction between parent and foreign affiliate, technological development of the affiliate by way of technology transfer can be much more intense (Birkinshaw and Hood, 1998).[11] This is what is referred to here as the 'dynamic, two-way knowledge transfer between foreign affiliate and parent investor'.

In case the foreign affiliate, however, matures with respect to its absorptive capacity without a corresponding upgrading of its position in the management relationship (autonomy), the institutional learning curve will remain relatively flatter, as will the intensity of internal technology transfer.[12] Hence, or the conceptualisation envisaged in this analysis, the criteria of both absorptive capacities and autonomy are needed. In terms of methodology, those two criteria define the determinants of internal technology transfer, and they in turn act as necessary conditions (without going as far as to hold that there is a pre-determinable and general relationship between capacities, autonomy, and effects). Hence, this indirect method allows only potentials to be determined, not the intensity of actual technology transfer.

The simultaneous use of the two criteria gives rise to a two-dimensional taxonomy of foreign affiliates that characterises their potentials for internal technology transfer (see Figure 4.1): the vertical axis locates foreign affiliates according to their management relationship with their parents: foreign affiliates operating under a dominant parent are located at the lower half of the taxonomy. In terms of technology transfer, the concept assumes that the potentials for static effects are particularly high where the foreign affiliate has a dominant parent that is willing and able to implement its own technology into the subsidiary. Foreign affiliates located at the top are more autonomous in the management of their own firms. Being autonomous, however, does not guarantee that the affiliate in fact significantly benefits from its foreign investor: only if it is able to adapt the foreign technology autonomously can the technology transfer be intense and of the more dynamic type. The ability of the affiliate to adapt the foreign technology is depicted on the horizontal axis. Affiliates located to the right have low absorptive capacities, whereas those located to the left share high such capacities.

The four stereotypical categories resulting from the two dimensions are assumed to be linked over time: in the lower right quadrant, the concept assumes foreign affiliates to be rather young and immature. They have low absorptive capacities, and more or less copy what they receive in terms of knowledge and technology without much adaptation. They also have a dominant parent and hence very little autonomy with respect to decisions over business functions in management. Here, the concept expects large potentials for a static, one-way internal technology transfer from parent to foreign affiliate. The foreign affiliate receives the parent's technology (dominant management position of parent), but is (so far) unable to contribute to its technological development by adapting the foreign technology to better suit its local environment.

If such affiliates do not develop their own absorptive capacities but at the same time gradually assume more responsibilities from their parent companies (that is, if they are 'left alone', see top right quadrant in the figure), then the concept assumes that potentials for internal technology

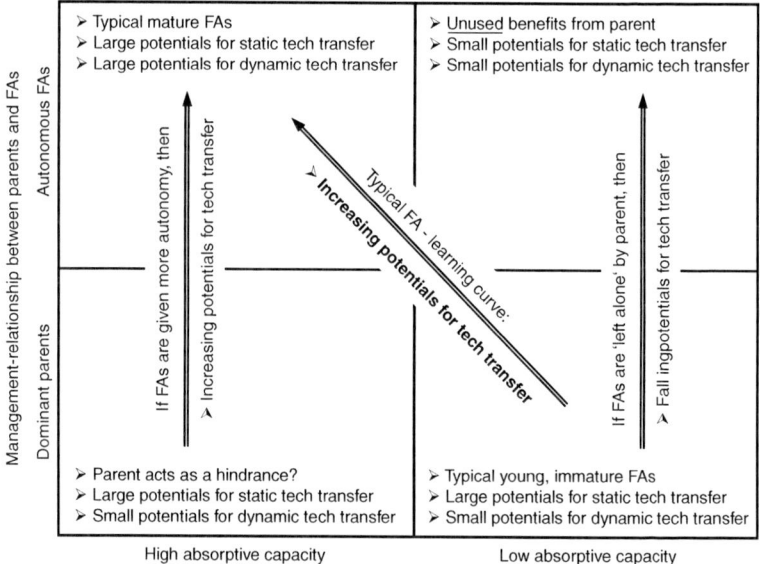

Figure 4.1   The conceptual taxonomy and potentials for technology transfer

transfer will even fall; the affiliate is not able to make best, or even much, use of the foreign technology that it receives.

Where foreign affiliates do develop their own absorptive capacities yet are not granted more autonomy from their parents (lower left quadrant), the concept characterises parent companies as a 'hindrance' to the development of a dynamic technology-generating and sharing network: their potentials for internal technology transfer are large but at the same time remain static. The parent forgoes potential benefits from the dynamic interaction with its maturing foreign affiliate. Here, the affiliate receives the parent's 'best practice', but is not, however, allowed to participate by adapting it to function efficiently in its own environment – despite its ability to do so.

Only in the case where the foreign affiliate simultaneously develops absorptive capacities AND is granted a more active role in the management relationship in the investor's network, will this network be able to generate potentials for dynamic, two-way internal knowledge transfer between affiliate and parent. Due to its high absorptive capacities, the foreign affiliate will make use of the parent's technology, and will be able to decide which technology to choose and how best to implement and adapt it (static effect). When reporting back to the parent, a dynamic process of technology transfer between parent and affiliate and back again can emerge.

In general, the assumption is that with foreign affiliates maturing over time, they will typically move from the bottom right quadrant to the upper left one. An affiliate that finds itself located in the lower left corner of the taxonomy can upgrade only if the corporate governance approach of the foreign investor changes, whereas for an affiliate in the top right corner the responsibility for reaping full benefit from having a foreign investor lies exclusively with itself and its ability to increase absorptive capacities.

Because CEE is a region that has a long industrial history and is already well advanced in the process of catching up with the West in terms of technology and factor prices, networks of parent investors and their affiliates can typically be expected to tend in this direction over time.

4.3   The data used for the analysis

Unfortunately, it is impossible to test the dynamics over time, because there is simply no time-series firm-specific data to proxy absorptive capacities and levels of autonomy. There is, however, data at the firm level to describe the location of foreign affiliates in some of the East

European host countries at one particular point of time – this is the IWH FDI Micro Database 2006–2007. The database contains information from 144 foreign affiliates in Croatia, 295 in East Germany, 110 in Poland, 220 in Romania, and 40 in Slovenia.

With respect to the management relationship of autonomy of foreign affiliates, the database gives the opinion of affiliates as to what degree either their own management or the foreign investor undertakes a set of seven business functions: (i) production and operational management; (ii) market research and marketing; (iii) basic and applied research; (iv) product development (product innovations new to the firm, but not necessarily new to the market); (v) process engineering (new or improved production or delivery methods, including changes in techniques, equipment and/or software); (vi) strategic management and planning; and (vii) investment projects and finance. The indices for each of these business functions are ranked from 1 if the business function is undertaken 'exclusively by the foreign owner network', that is, no autonomy, to 2 for 'mainly foreign investor network', to 3 for 'mainly the foreign affiliate', and 4 for 'only the foreign affiliate'. The variables are hence discrete along an ordinal Likert scale ranging from 1 to 4 with increasing levels of autonomy.

For the proxy of absorptive capacities, the analysis uses own R&D, as is so often used in the literature (Cohen and Levinthal, 1989, 1990). The database contains information on R&D or innovation-related expenditure in per cent of total sales 2005 and the share of R&D personnel in 2005. This information was directly interrogated from the managers of the foreign affiliates and cross-checked with firm-financial databases where this was possible (using the sources for the total population). The two sets of information on R&D employment and R&D expenditure are amalgamated into a common indicator by way of multiplication, that is, just like in an interaction term: this way, absorptive capacity is high where both expenditure and personnel for R&D are high simultaneously, and vice versa. Where firms employ a large number of R&D personnel yet do not spend much in terms of R&D expenditure, or where large expenditure is not matched by a large share of specialised R&D personnel, absorptive capacities are assumed to be less well developed.

Furthermore, because both knowledge-related proxies can be expected to be quite industry-specific, with large variations between industries as a typical feature, the analysis uses firm-specific deviations of the 'composite indicator' from the average of the indicator in the respective two-digit NACE industry. This yields positive and negative firm-specific deviations, and their magnitude reflects the extent to which the foreign affiliate has a higher or lower absorptive capacity in comparison to what is typical in the firm's own industry on average in the CEECs of the database.

4.4 Characterisation of foreign affiliates in CEE within the two-dimensional concept

The theoretical version had been developed to lend itself to an empirical representation of foreign affiliates in Central East Europe by use of the IWH FDI Micro Database. This representation should provide a picture of which of the countries appear to contain the largest potentials for a dynamic, two-way internal technology transfer via FDI, given their individual endowment with different kinds of foreign affiliates.

In the analysis, the focus is on differences between countries, because the objective of this study is to assess country-specific potentials for technology transfer via FDI (see RQ(4_2). It is hence implicitly assumed that country differences are more important than industry differences: in fact, the standard deviation of industry-specific means (13 two-digit NACE) turn out to be slightly lower (0.145) than that of country-specific means (0.150). In an earlier analysis made with data from the pilot study generated between 2002 and 2003, foreign affiliates were located in groups of industries according to their technology intensity (OECD-classification,[13] see Hatzichronoglou, 1997 and OECD, 1999). The results of this analysis showed that foreign affiliates in rather low-tech industries appeared to be generally more autonomous than those in industries with, usually, higher technologies. And affiliates in low-tech industries on average appeared to have rather higher levels of absorptive capacities (see Hamar and Stephan, 2005). Although the method has important weaknesses, the results do suggest that the more technology-intensive the foreign investors, the warier they become of granting too much autonomy. That corresponds to the findings of related case-study research on CEE (for example Dyker et al., 2006).

4.4.1 Autonomy in business functions

Across the whole sample, the average over all autonomy indices amounts to 2.77 (with a standard deviation of 0.82, n = 716). Not surprisingly, the highest levels of autonomy are recorded for production-related business functions (i and v). The lowest levels are recorded for (vii) investment projects and finance, (iii) basic and applied research, and (vi) strategic management and planning, where foreign investors are more likely to retain control.

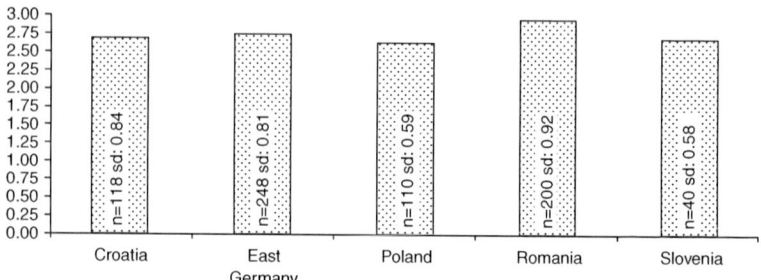

Figure 4.2   Average levels of autonomy in all seven technology-related business functions of foreign affiliates across countries

Note: Unweighted averages of autonomy levels (4 for maximum autonomy, 1 for minimum autonomy) in all seven business functions. n is the number of firms that have provided information on all business functions. sd denotes standard deviation.

Source: IWH FDI Micro Database 2006–2007.

In a host country comparison (see Figure 4.2), managers of Romanian foreign affiliates feel the most autonomous on average, which is a result of high autonomy levels in mainly technology-related business functions. Romania is followed by East Germany by a larger gap, whilst Croatia and Slovenia follow closely, and finally Poland with the lowest level of autonomy: here either foreign investors are in fact less willing to allow their affiliates to take their fate in their own hands, or else the expectations of affiliate managers in these two countries are somewhat more demanding with respect to autonomy.

4.4.2   Absorptive capacities

To provide an empirical picture of absorptive capacities at the country level, negative and positive deviations of all foreign subsidiaries are averaged to provide a picture of the typical endowment of each country with absorptive capacities amongst firms (see Figure 4.3). For the whole sample, the average over all deviations is of course 0. The country ranking here differs from the ranking of autonomy levels: East German foreign affiliates have the highest mean of above-industry-average absorptive capacities, followed by Romanian and Croatian affiliates, whereas Slovenian and Polish firms have the lowest levels (highest negative deviations of industry averages).

It is in fact not a surprising result that East German foreign affiliates lead the ranks:[14] after all, the proxy measures R&D-related indicators, and here, East German firms can be expected to be technologically more active than those of the other countries of the database. At the very least, this reflects the typical association of levels of economic development, measured in GDP per capita (or average technology level of the country, measured in aggregate productivity) and the intensity of technological activity amongst firms. Related research likewise suggests that the East German National Innovation System is by far the most developed amongst all transition economies or regions, and here in particular the state of transformation of research institutions (universities, public and private research institutes, like the Fraunhofer, the Max Planck, the Leibnitz list) (see for example Günther et al., 2008b). In the case of CEE, research institutions have typically emerged from national academies of sciences, and today often remain insufficiently transformed to be able to play an active part in the regional innovation systems (McGowan et al., 2004; Meske, 2004; von Tunzelmann, 2004; von Tunzelmann et al., 2010). Their history of economic planning and their orientation in external economic relations on predominantly the CMEA (von Tunzelmann, 2004) still make most research institutes inferior to their western research counterparts today (in the words of von Tunzelmann et al., 2010, they suffer from network failures with regard to industrial R&D).[15]

The surprisingly high level of Romania may be a statistical phenomenon: what counts as R&D expenditure and R&D personnel depends on each country's formal and informal definitions, and here Romania may be more lenient in allowing expenditure and personnel to be tagged as R&D.

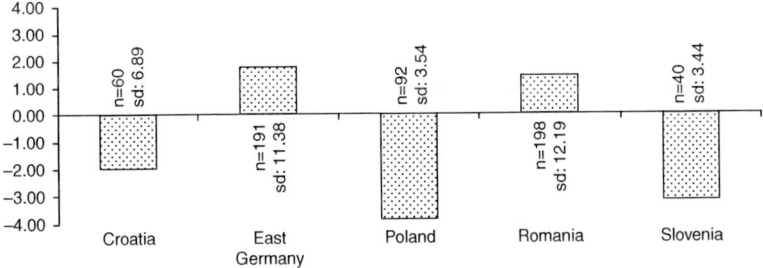

Figure 4.3 Levels of absorptive capacities per host country

Note: Means of positive and negative deviations from industry averages of the proxy of all firms. n is the number of firms that have provided information on R&D employment and R&D expenditure. sd denotes standard deviation.

Source: IWH FDI Micro Database 2006–2007.

4.4.3   The location of foreign affiliates in the empirical representation of the taxonomy

The proxies for autonomy and absorptive capacities are then used to locate all foreign affiliates into an empirical representation of the conceptual taxonomy. On the vertical axis, the levels of autonomy for each firm are depicted, the firm-specific deviations of industry-average absorptive capacities on the horizontal axis.[16] Each little x represents one of the 510 foreign affiliates of the sample that answered the questions that were necessary to construct the indicators (see Figure 4.4). The graphical interpretation is difficult, because several firms may have the same numerical combination of both indicators. In that case, the xs will overlap and it will not be possible to discern that this represents more than one firm. The graphs still provide us with a tentative indication as to how the countries compare in terms of endowment with different kinds of foreign affiliates (answering RQ(4_2) and is completed by the numerical information provided in Table 4.1. From this, we attempt to

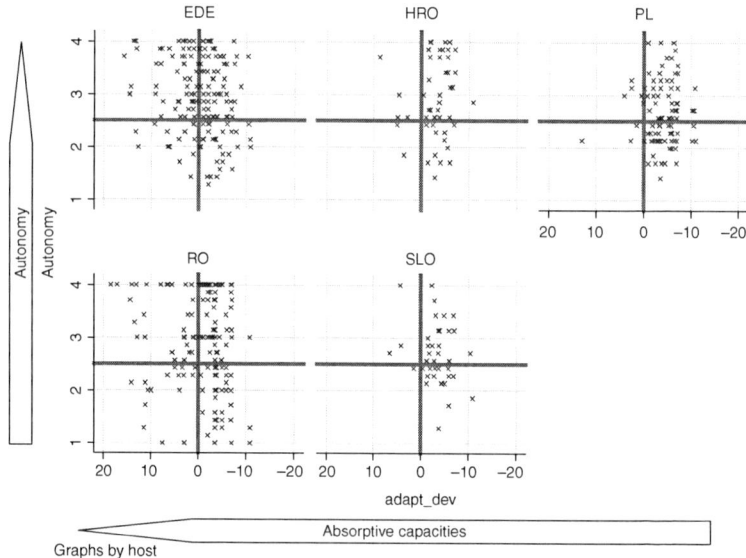

Figure 4.4   Graphical representation of empirical taxonomy

Note: EDE stands for East Germany, HRO for Croatia, PL for Poland, RO for Romania, and SLO for Slovenia.

Source: IWH FDI Micro Database 2006–2007.

Table 4.1 Shares of foreign affiliates allocated into the four quadrants of the taxonomy

"Mature FAs"			"Left-alone FAs"
East Germany	33.3%	East Germany	39.5%
Croatia	8.2%	Croatia	63.3%
Poland	5.4%	Poland	47.8%
Romania	21.0%	Romania	44.3%
Slovenia	7.5%	Slovenia	50.0%
Total	19.8%	Total	45.7%
"Parent as hindrance"			**"Immature FAs"**
East Germany	8.6%	East Germany	18.5%
Croatia	4.1%	Croatia	24.5%
Poland	3.3%	Poland	43.5%
Romania	10.2%	Romania	24.6%
Slovenia	2.5%	Slovenia	40.0%
Total	7.2%	Total	27.3%

Note: Percentages are calculated as shares of foreign affiliates of one country in the total number of firms of that country, not of the totality of foreign affiliates. Hence, all numbers for each country add up to 100 per cent (notwithstanding rounding errors).

Source: IWH FDI Micro Database 2006–2007.

infer the host economies' respective potentials for internal technology transfer between investor and affiliate.

At first glance, FDI into the region appears to be somewhat denser on the right-hand sides of the taxonomies, that is, a bias on similarly low negative deviations from the respective NACE two-digit industry averages. On the left-hand side of the taxonomies, absorptive capacities are more dispersed, with some firms already achieving higher levels. In terms of numbers (Table 4.1) and for all countries together, the largest share of foreign affiliates (45.7 per cent) is located in the top right quadrant, denoting 'left-alone foreign affiliates' with neither large potentials for a two-way internal technology transfer between parent and foreign affiliate nor a large potentials for a static one-way transfer from parent to foreign affiliate. 27.3 per cent are located in the lower right quadrant, that is 'immature foreign affiliates' with only little potentials for a dynamic two-way technology transfer yet large potentials for the static one-way kind of technology transfer.

In only 7.2 per cent of cases would the analysis assume that the parent firm acts as something of a hindrance to the technological development of the foreign affiliate by not granting the kind of autonomy that the level of absorptive capacity of the firm would warrant. Nearly 20 per cent of all firms, however, are located in the region in which the analysis characterises firms as containing large potentials for both the static and the dynamic kinds of technology transfer.

At the country level, using the geographical centre of all foreign affiliates in the space of the taxonomy, results suggest that East German foreign affiliates (and possibly Romanian) have probably the largest potentials for a dynamic technology transfer (top left), and Croatian, Polish, and Slovenian foreign affiliates, probably the lowest potentials (top right). In terms of shares of foreign affiliates (Table 4.1), 33.3 per cent of all East German foreign affiliates are located in the top left quadrant. Whilst in Romania a large share (20 per cent) is also located in this quadrant, all other countries assessed here have very low shares in this quadrant, and hence suggest very low potentials for a two-way dynamic technology transfer (Croatia: 8.2 per cent, Slovenia: 7.5 per cent, and Poland with only 5.4 per cent). The top right quadrant has high shares mainly in Croatia (63.3 per cent), but also in Slovenia (50 per cent) and Poland (47.8 per cent). This is the location where the concept would expect rather a low potentials for technology transfer. Yet Poland has an almost equally large share of firms located in the bottom right quadrant, where foreign affiliates are assumed to enjoy the possible prospect of being able to move up the learning curve.

The results hence suggest a ranking of countries in terms of potentials for technology transfer, based on their endowment with different kinds of foreign affiliates with respect to comparative levels of autonomy and comparative absorptive capacities. Such a ranking would probably be headed by East Germany and followed by Romania. If a location in the top right quadrant can be assumed to signify a half-way stage towards the top left quadrant where potentials for dynamic technology transfer are highest, then the ranking would proceed with Croatia and Slovenia. This assumes that those countries' subsidiaries are in a position to improve their potentials for technology transfer by increasing their adaptive capacities, having already achieved some autonomy from their parent investor companies. Poland would then assume a lower position in such a ranking; a large share of Polish foreign affiliates may be characterised as being near the start of the typical foreign affiliate-learning curve that was characterised in the conceptual taxonomy.

4.5 Summary of main results

The literature review in the chapter finds that internal technology transfer via FDI into CEECs in general can be expected to be significant (and much larger than external effects). The main determinants of an intense internal technology transfer that goes in both directions (RQ4_1) are found to be the industry that foreign affiliates belong to (including the extent of competition not least due to the extent of external penetration of the market); the extent of absorptive capacities of foreign affiliates with a view on making best use of alien technology in their own environment (which depends on: the size of the technology gap between home and host country or institution, the two faces of R&D, state support for R&D, the ability to learn, human capital in general); the strategic motive of the foreign investor; the extent of tacitness of knowledge and technology the investors are endowed with; the shape of the management relationship within the network of the investor – the autonomy of the affiliate vis-à-vis its investor; and finally the state of development of formal institutions (as for example the enforcement level of the IPR regime).

The analysis uses those determinants to assess the potentials of host economies in CEE to benefit from their foreign investments in terms of a dynamic, two-way internal technology transfer. Of the many determinants found in the literature, three were singled out for the subsequent analysis (industry, absorptive capacities, autonomy). It is proposed that applying those to foreign affiliates in CEECs is sufficient to produce an insightful and valuable representation of the endowment of CEECs with foreign affiliates of different characteristics according to those three determinants (RQ4_2).

A conceptual taxonomy is developed to determine the potentials for technology transfer by assessing the endowment of the countries with foreign affiliates, characterised along the two criteria of absorptive capacities and autonomy (RQ4_3). A ranking follows from the empirical representation of this taxonomy, and this suggests – assuming representativeness – that East Germany contains the largest potentials for dynamic, two-way internal technology transfer between foreign parent and its affiliate. East Germany is followed by Romania, but with a larger share of affiliates where the parent investor may be considered a hindrance to the further development of the affiliate, as well as a larger share of affiliates that have yet to increase their own technological abilities. The latter characterisation also applies to Croatian and Slovenian affiliates. Poland may be interpreted to assume the last position in the ranking, mainly because its foreign affiliates can predominantly be characterised as technologically immature, due to comparatively low absorptive capacities and at the same time because they remain quite dependent on the decisions of their foreign investors. This lowest rank, however, is also attached to the possible development of affiliates along the typical learning curve.

Those results, however, have to be read with due care: first, the answers of foreign affiliates with respect to autonomy are subjective – and possibly also country-specific due to differences in mentalities. Furthermore, the large size of deviations around the geographical centre need to be taken into account when interpreting the results where they present averages: for autonomy, standard deviations are particularly high in Romania, but the levels of standard deviations in per cent of the means here are nothing like the levels for absorptive capacity in all countries, and here in particular for Romania again and East Germany. Some of this effect is due to the particular construction of the index as an interaction term.

5
Central and East European Innovation Systems as Knowledge Sources for Foreign Affiliates' Own Technological Activity in CEE

Potential technological effects via inward FDI not only emerge from direct, internal technology transfer. The contribution to technological catch-up development is also expected to emanate from the diffusion of knowledge and technology between the host economy and its foreign affiliates (the 'external' part of technology transfer and spillovers). This technology transfer may assume the characteristics of a 'transfer' of knowledge and technology (that is, intentional, enshrined in contracts, and not giving rise to externalities) and/or can rather be of a 'spillover' type, where this diffusion is unintentional and not regulated in contracts, and hence gives rise to externalities. In line with the conceptualisation of research in this work, this distinction does not play a role in the analysis of this chapter (nor in the following one, Chapter 6). The focus in this chapter is rather on the conditions of positive technological effects from the host innovation system for the foreign affiliates hosted by them. Those conditions are deduced from theoretical (or conceptual) and empirical research, and give rise to a set of hypotheses on what drives the role of the host innovation system for technological activity of the foreign affiliate.[1] The empirical analysis presented here tests whether empirical support for those hypotheses can be found in the particular case of CEECs. In terms of theories, the analysis mainly draws from Cantwell's technology accumulation approach within the body of international business literature, assuming that there is a "complex dynamic interaction of the ownership advantage of groups of firms and the locational advantages of the sites in which they produce" (Cantwell, 1989, p. 207). This view has its firm roots in the innovation literature that holds that innovation is typically a result of a recombination of existing knowledge (Kogut and Zander, 1992; Nonaka, 1994). From an organisational perspective, technological innovation emerges typically not only from sources within the boundaries of the firm (intra-mural), but also from outside sources. This stresses the important role of the interfaces between firms, universities, research laboratories, suppliers, customers, and potentially others (Powell et al., 1996). The generation of new knowledge and ideas often requires some extent of collective efforts, involving different stakeholders in cooperative actions (Chesbrough, 2003).

It is an important implication of the use of multiple sources for the generation of new knowledge and technology that this may lead to (i) increases in technological opportunities (see for example Klevorick et al., 1995) as well as (ii) complementarities and synergies between knowledge sources (see for example Leiponen and Helfat, 2004). Whereas academic research may provide important knowledge for industrial innovative activity, this is typically not of the more applied sort that easily lends itself to commercial exploitation directly, which is what firms tend to do (Mansfield, 1991; Pavitt, 1998). In contrast, agreements between commercial firms tend to focus more on products closer to their commercial use (Arora and Gambardella, 1990). Users of innovations – firms or individual consumers – are potentially able to provide knowledge of relevance to the supplier, where it is concerned with, for example, problems associated with the innovation and/or suggestions for further modifications (von Hippel, 1976, 2005). The opposite direction is also conceivable: suppliers may provide producers with knowledge regarding the product or service specifications as well as regarding the management of supplies (processes, logistics, etc.). These insights from the innovation literature provide very useful and highly complementary perspectives for research into what the international business literature has to hold on the interaction between FDI affiliates and the innovation systems in their host economies.

This sets the agenda for the two research questions of this chapter:

RQ(5_1)   Under what conditions do foreign affiliates draw on local knowledge sources?

RQ(5_2)   What are the important determinants of a positive role played by CEECs' host innovation systems for the technological activity of their local foreign affiliates?

Before the first research question is considered, this chapter starts with a precondition to R&D and innovation activities within foreign affiliates.

The review of the literature on internationalisation of technological activity in TNCs strongly suggests that the empirical evidence of widespread technological activity amongst foreign affiliates in CEECs can be rationally explained. Proceeding from there, answers to the first research question are generated from a targeted review of the international business and international management bodies of literature. This includes a review of the current state of the art in researching the internationalisation of technological activity (R&D and innovation) within TNCs. The following review assumes the perspective of the FDI project, and focuses on its potential strategies and technological capabilities. The review from the perspective of the host innovation system individually considers its three main actors: suppliers in upstream industries, customers in downstream industries (including end users), and scientific research institutions. Those actors are assessed in terms of their technological capabilities and the effects of spatial proximity between those actors and the foreign affiliates.

The second research question is addressed by way of empirical analysis. This tests a set of hypotheses on the determinants of a positive role fulfilled by the host innovation systems in CEECs.

5.1 Internationalisation of technological activity in transnational corporations

One of the important channels of technological development via FDI is the link provided by foreign affiliates to global technology and innovation networks. In most industries, the technological frontier is defined by large TNCs, as they are major players in the use and development of knowledge and technology on a global scale. In empirical terms, TNCs account for the majority of business expenditure on R&D; they dominate new patents, and often turn out to be role models for new management and organisation methods (see UNCTAD, 2005, p. 99). R&D can be divided into basic and applied R&D, whereby TNCs often organise basic research in a more centralised manner, probably at the headquarters or a developed host economy with low or negative productivity gaps. Applied research, however, often follows the aim of adapting a product, process or service to the local host economy environment, and is hence more easily organised in a decentralised way, and may, profitably for the whole TNC network, involve the home countries' innovation system (for example Kvinge, 2004).

In the past, TNCs tended to locate R&D in their headquarters (which also tended to be in developed countries) or else in foreign affiliates located in developed countries. Where R&D finance and mandates were in fact granted to affiliates in developing or emerging countries, their main task was typically to adapt products, services or processes to local conditions. This behaviour was perfectly rational: R&D that requires strict control and where transaction costs (in R&D, this is due to the often tacit nature of knowledge) are high (Buckley and Casson, 1976) and where ownership advantages involve proprietary knowledge and information that is expensive to produce yet easy to replicate (Dunning, 1989), R&D requires strict control and hence would have been typically allocated to the headquarters. Furthermore, technological activity tends to be embedded mainly in the home environment (Narula, 2002) and one of the pertinent characteristics of efficient technological activity is its need for internal cohesion within the firm (Blanc and Sierra, 1999; Zanfei, 2000). The 'non-globalisation' case for R&D "reflects the linguistic and geographic constraints imposed by person-embodied exchanges and transfers of tacit knowledge" (Patel and Pavitt, 2000, p. 218)[2].

This, however, began to change during the 1990s and the early 2000s. Transaction costs have come down due to modern communication instruments; travel costs and travel times have come down significantly; the legal frameworks that help to protect sensitive IPR and their enforcement have improved markedly in the most recent past (see the enactment of minimum standards through the TRIPS agreements as WTO rules since 1995). Moreover, no country can provide all technology and knowledge, or the ability to generate such, on all technology fronts (Criscuelo et al., 2002, p. 19). TNCs' largest competitive advantage is their ability to shop around the world to find what is needed at the best quality, cheapest prices etc. In terms of R&D, the same applies: they are able to exploit diverse international forms of knowledge, ideas, tastes, institutions supporting innovation etc. on offer in different locations, and they are hence able to overcome the restrictions of technological specialisation within their home regions (see for example Cantwell and Piscitello, 2005; Frost, 2001). This is also reflective of Kogut and Zander's (1993) view that TNCs develop comparative advantages by way of specialisation in the capability to transfer tacit knowledge within the firm, which then becomes a "repository of social knowledge that structures cooperative action" (ibid., p. 627).

The traditional view on advantages of centralised technological activity in home countries is becoming less relevant, and research has increasingly focused on the advantages of decentralisation of technological activities (see for example Pearce and Singh, 1992; Howells and Wood, 1993; Miller, 1994). This decentralisation offers linkages with local markets, suppliers and clients. It also offers the exploitation of potential technological fields of excellence in host economies (see for example Dunning and Wymbs, 1999; von Zedtwitz and Gassman, 2002; Cantwell and Iammarino, 2003; Cantwell, 1992, 1993).

In a more recent trend during the last one or two decades, TNCs appear to more readily venture into regional dispersion of industry-specific core technologies as well (Cantwell and Santangelo, 2000). Not only has the overall degree of internationalisation of R&D been increasing (see for example UNCTAD, 2004, focusing on services and using a large survey; Ambos, 2005, for Germany; Edler and Polt, 2008, in a special journal issue on policy implications). R&D is also being allocated more often in developing countries, starting with Asia, with some R&D already at par with work undertaken in the developed economies (see UNCTAD, 2005, p. 100). Cross-border technology transfer not only occurs via technology-intensive trade but also through international licensing and patents; international cooperation (measured by patents with foreign co-inventors) is on the rise, technology-related direct foreign investment has also risen (Edler and Polt, 2008, pp. 332–33). Today, differentiated "global innovation networks" (Sachwald, 2008) are taking a foothold in the international technological world in general, and in emerging and developing countries in particular. It is important to note, however, that foreign affiliates have not internationalised their innovative activity in proportion to the growth of their overall production activities (see for example Zanfei, 2000; Patel and Pavitt, 1999).

For TNCs today, the most important locations of technological activity are neither the low- or least-developed countries, nor the industrialised countries. Rather, the TNCs are on the lookout for localised knowledge and technology in emerging markets, and this also includes CEECs. In those countries, a large share or even the bulk of innovative output originates from foreign affiliates in CEE, whereas innovation activities amongst domestic enterprises in CEECs are less intense: innovation systems in CEE are increasingly dominated by foreign affiliates (see for example Halpern and Muraközy, 2009). Jindra and Rojec (2012) find that the share of foreign affiliates in domestic business R&D expenditures has increased substantially, in a comparison between the periods 2000–2003 and 1995–1999 in Latvia, Estonia, and Hungary, whilst the increase was less pronounced in Slovenia, Slovakia, Poland, and the Czech Republic. As early as 2003, the share were already as high as 62.5 per cent in Hungary and 46.6 per cent in the Czech Republic, although in Poland and Slovakia, the shares were still only around 19 per cent (UNCTAD, 2005, p. 127). These shares can be expected to have risen even further since then.

Innovations in those countries are often carried out on the shop floor rather than in science labs, and even if they do not often push the frontiers of scientific knowledge, they are commercially important. Whether foreign affiliates in host economies with a productivity gap *vis-à-vis* the investors' home countries are mandated to research and develop within the network of their foreign investors (much along the lines of 'knowledge as a public good within the firm': Buckley and Casson, 1976, p. 35), will not only depend on the characteristics of the foreign affiliate itself alone; it will also depend on the capability of the innovation system of the host economy with its multiple actors and institutions to provide a unique knowledge base that can be beneficial to the foreign affiliate and its foreign investor network.

There is very little general literature that deals explicitly with the determinants of technological activity of foreign affiliates, or from the perspective of the headquarters: determinants of internationalisation of technological activity within TNCs. In some studies, foreign affiliates' technological capabilities and intensities of innovative activities are related to the availability of TNC-internal innovation networks (Fratocchi and Holm, 1998, in an analysis of 'centres of excellence'; Birkinshaw and Ridderstrale, 1999, in a study of 'subsidiary initiatives'; Pearce and Papanastassiou, 1999, on UK-based research laboratories; Andersson et al., 2002; Yamin and Otto, 2004). With regard to CEECs, Kokko and Kravtsova (2008) focus on innovative capability in foreign affiliates in four transition countries (Estonia, Hungary, Poland, Slovenia) and suggest three sets of determinants: (i) the role of the affiliate in the foreign investor's network; (ii) affiliate characteristics such as size, age, industry; and (iii) host country and host industry characteristics, as for example the level of development, competitive pressure etc. They find that innovative capabilities increase with the level of independence of the foreign affiliate from its foreign investor (see the autonomy issue assessed in Chapter 4). In CEECs, it is not foreign affiliates in the usually high-tech industries that exhibit particularly large innovative capabilities, but rather those in industries at the lower end of the technological spectrum: even if an FDI project is in a typically high-tech industry (such as pharmaceuticals), then R&D at the foreign affiliate level in CEECs is – if it in fact exists – often of a lower commercial value.[3] Next to the findings on the autonomy issue (as described in Chapter 4), Damijan et al. (2011) also find that access to the knowledge base of the foreign parent markedly increases the innovative intensity of a foreign affiliate, whilst a market-seeking motive is associated with lower product innovativeness. Patent citation analyses suggests that whilst foreign firms are indeed important sources for patenting, the sources of knowledge leading to patents do not typically reside in the host economies or even foreign affiliate's own R&D efforts. Rather, patents tend to cite already existing technology of the foreign investor network (Manea and Pearce, 2006). In CEECs, national research institutions, like academies of science that originate from the former socialist era, appear to be of less importance, which is related to the finding by von Tunzelmann (2004) that those institutions still remain rather inefficient in CEE (East Germany may be an exception here; the issue of post-socialist academies of science is discussed later in this chapter, where the capabilities of host innovation systems in CEE are reviewed). Likewise, domestic firms are found to be insufficiently capable in terms of technology development (see Dyker, 2006; Dyker et al., 2006; Varblane et al., 2007). Günther et al. (2009) find that whilst the majority of foreign affiliates in CEE are technologically active, their technological capability is largely detached from the host country environment.

In general, however, the question of whether the causality runs from foreign affiliates' technological capability and activity to external technological sourcing or vice versa is an unresolved matter in the literature, and both directions are perfectly plausible. This is why the issue of determinants of technological activity of foreign affiliates is not further analysed empirically by use the IWH FDI Micro Database. Rather, the subsequent analysis only includes those firms in the analysis that are in fact technologically active, and assesses the determinants that make such foreign affiliates draw on the host economy knowledge base.

5.2 The current state of the art in research

The analysis of conditions under which foreign affiliates draw from local knowledge sources has to consider and distinguish between strategies and capabilities of the foreign investor network and the capabilities of the host economy innovation system to supply and make available knowledge and technology valuable to the foreign investor network.

Existing conceptual or theoretical insights into the determinants of use of the local knowledge base by foreign affiliates draw upon both the literature on innovation, and here with a systemic perspective on internal and external networks to take account of the network character of TNCs. And they draw on the international business literature on TNCs, here with a focus on the technological development of the partners within the MNC network. Both strands are used to find out under what conditions foreign affiliates make use of the innovation systems of CEECs hosting FDI. The first set of conditions takes the perspective of the foreign investment network, and the other set assesses what conditions have to pertain within the host innovation systems in CEE.

5.2.1   Foreign investor strategies and capabilities

With a view on the role of the foreign investor, the distinction between the strategies of **'home-base exploiting'** on the one hand and **'home-base augmenting'** (Kuemmerle, 1997) on the other is useful for the analysis of the conditions under which foreign affiliates interact with the local host economy. This dichotomy in the conceptualisation of strategies reappears in several different connotations in the literature, and still largely describes the same idea: 'home-base exploiting' corresponds to 'competence exploiting' behaviour, whereas 'home-base augmenting' corresponds to 'competence augmenting' behaviour (Cantwell and Piscitello, 1999; Cantwell and Mudambi, 2005).

Probably the earliest account of such a dichotomy and often referenced as source of the idea is March's (1991) distinction between 'exploitation' and 'exploration': the former is characterised by "the refinement and extension of existing competences, technologies, and paradigms" (March, 1991, p. 85). Here, the scope of knowledge searching is more likely to reside in the routines and problem-solving heuristics within the MNC's established knowledge base. Foreign investors concentrate on using the stock of knowledge and technology within their MNC network and home base, with adjustments exclusively to adapt the product or service to local host economy's individual characteristics (refinement and adaptation). The knowledge base of the host innovation system is of less importance here; technological interaction between the foreign affiliate and the host economy is weak, and the benefits for the host economy of having a foreign investor in terms of technological upgrading are rather limited. This parallels the contention of Frost (2001), where the "foreign subsidiary's innovation is adaptive in nature" (p. 105) and "will be more likely to draw upon technical ideas originating in the home country" and "less likely to draw upon technical ideas originating in the host country" (ibid.).

March's exploration strategy, on the other hand, entails "experimentation with new alternatives" (March, 1991, p. 85) and "organizationally more distant from the locus of action and adaption" (ibid., p. 73). This corresponds to the home-base augmenting strategy, and it is this strategy that relies on the local knowledge base of the host country. Where the foreign investor seeks knowledge and technology that is complementary to its own and that can be beneficial for the whole foreign investor network, interaction with the hosting innovation system will be more intense. The technological benefits for the host economy can be larger. According to Criscuelo et al. (2002), this however depends on the efficiency with which the foreign investor can make use of the complementarity of resources. Because of the tacit (or 'sticky', von Hippel, 1994) nature of knowledge and technology, for the home-base augmenting strategy to work the foreign affiliate not only has to be in close proximity to both relevant actors in the host innovation system, but also, due to the problems associated with managing cross-border R&D activities often in multiple locations around the globe, internal proximity between all R&D facilities of the MNC network is needed just as much.

Cantwell and Mudambi (2005) analyse R&D intensities in UK foreign affiliates of non-UK multinational enterprises and extend the literature on those two foreign investor strategies. They assume the perspective of the foreign affiliate: their drivers of affiliate R&D distinguish between 'competence exploiting' (demand driven) and 'competence-creating' (supply driven) strategies of the foreign affiliate itself, not the strategies of the foreign investor. They likewise find that "the degree to which subsidiaries are separately granted strategic independence [...] positively influence R&D in competence-creating subsidiaries, but not in other kinds of subsidiary." (ibid., p. 1125).

In the case of both groups of strategies, Dunning's locational advantages are important: for the home-base exploiting strategy, it is the demand in the host location that characterises the locational advantage of the host location. Differences in demand give rise to product and process differentiations, the development of which give rise to technological development. And it is the two factors of supply of new ideas and complementary knowledge and technology available in the host location that define Dunning's locational advantages for foreign investors with a home-base augmenting strategy. The concept of the augmenting strategy leads to an important aspect in Cantwell's theory of technological accumulation: rather than passively taking advantage of given locational advantages, the foreign investor actively generates or improves locational advantages by way of allocating technological activity to the host location. The host location innovation system is treated as an integral and active part of the MNC's internationalisation and technological strategy. The theory of technological accumulation explicitly distinguishes between the technological role of the internal network (potentially giving rise to internal technology transfer between parent and affiliate) and external networks (granting the foreign investor access to localised knowledge and technology that exists or can be developed in the host region). The foreign investor who is particularly successful in activating the potentials that lie with the host region innovation system generates competitive advantages. Here, it is not an *ex-ante* existing market power due to market failure that drives internationalisation. This is a concept that can be firmly rooted in the dynamic, evolutionary view based on learning (see Nelson and Winter, 1982).

This concept of those two strategies is similar to the distinction between **local market orientation** and **export orientation:** Cantwell and Mudambi (2005) suggest that a local market orientation in sales may lead to the customisation of products and processes to the particular local market environments (product differentiation and process adaptations/innovations). Local market orientation can be assumed to be related to more intense collaboration between the foreign affiliate and local customers. Export orientation, however, may involve less interaction with the host innovation system. This becomes particularly apparent in the extreme case of OPT: here, rather little interaction with the host economy is possible (see the discussion in Chapter 4). The same mechanisms with their resulting relationships between the host innovation system and technological activity of the foreign affiliate should be expected to apply to the relationship with local suppliers, even if this had been to some degree overlooked in the literature so far (see Günther et al., 2008b). This idea is referred to in the summary of empirical literature in Hoekman and Smarzinska (2006) and closely related to the research carried out by Giroud (2007) on MNCs in Malaysia and Vietnam: foreign affiliates should benefit from technological spillovers through backward vertical linkages, if the benefits from vertical technology transfer are assumed to go in either direction.

A further aspect of foreign investor strategies is that of corporate control that the foreign investor applies on its affiliate, or the degree of centralisation of decision making, or foreign affiliate autonomy, as discussed in Chapter 4. Here, the autonomy issue again plays a role, not inasmuch as it complements adaptive abilities to give rise to an intense, reciprocal internal technology transfer, but rather, autonomy is treated here as a determinant of the ability of foreign affiliates to reach beyond their foreign investors' networks as a source of knowledge and technology for their own R&D and innovative activities and to generate own initiative to improve their own technological performance. In an empirical analysis testing how Canadian, Scottish, and Swedish foreign affiliates develop, Birkinshaw et al. (1998) find that the foreign affiliate autonomy and the entrepreneurial culture in the foreign affiliate were the key drivers of foreign affiliate initiative.[4] Birkinshaw and Hood (1997), in an exploratory study on mature, developed foreign affiliates in Canada and Scotland, find similar results: the greater the autonomy of the foreign affiliate the better its ability to form external network linkages with other companies and institutions in its own local environment. Likewise, Egelhoof et al. (1998) find that a strong assignment role for headquarters, that nevertheless encourages subsidiary initiative at the same time, is beneficial to the performance of foreign affiliates.

Another theoretical concept in the international business literature that has important implications for the first research question in this chapter (RQ5_1) as to under what conditions foreign affiliates draw on local knowledge sources, is the one on MNC evolution and subsidiary development, or 'developmental subsidiary', as also discussed in Chapter 4.[5] This builds on the concept of 'subsidiary roles', where foreign affiliates may add increasing value to the MNC and its network (see mainly Bartlett and Ghoshal, 1986, 1989, 2004; Bartlett and Ghoshal and Birkinshaw, 2004; Birkinshaw and Hood, 1997; Birkinshaw and Morrison, 1996; Ghoshal and Nohria, 1989; Gupta and Govindarajan, 1994; Hedlund, 1986; Jarillo and Martinez, 1990; Roth and Morrison, 1992). The source of firm-specific advantage for the MNC is distinct from the original assumption that the foreign owner supplies a proprietary or ownership advantage (centrality à *la* Hymer, 1976, or Dunning, 1980, 1988): it pays tribute to the empirical fact that foreign affiliates can play an important part in the creation and maintenance of MNC-specific advantages. Foreign affiliates can contribute to or lead innovation projects (Bartlett and Ghoshal, 1986), they can provide resources to the rest of the MNC network (Gupta and Govindarajan, 1994), and foreign affiliates may have the mandate to develop and produce certain product lines for the whole MNC network (Roth and Morrison, 1992).

The concept of subsidiary development further extends these insights by assessing how foreign affiliates become able to contribute to the firm-specific advantages of the MNC network. Here, the focus is on the foreign affiliate's ability to make valuable use of the host economy's regional innovation system: Birkinshaw et al. (1998) suggest three "perspectives"[6] (p. 222–3): the first, "environmental determinism", describes the notion that the host environment determines the potential of a foreign affiliate to actively contribute to the MNC network (in general: Ghoshal and Nohria, 1989; Westney, 1994; if the host economy is of strategic importance for the MNC network: Bartlett and Ghoshal, 1986; as an effect of dynamic local competitors, suppliers, and customers: Porter, 1990; with a view on policies, direct and indirect: Birkinshaw and Hood, 1997). The second perspective is on "head office assignment", where the foreign investor's headquarters controls and coordinates its foreign affiliates, and this according to the assumed endowment of host countries with locational advantages (Birkinshaw and Morrison, 1996; Gupta and Govindarajan, 1994; Roth and Morrison, 1992; distinguishing between investment and divestment, spin-offs: Birkinshaw and Hood, 1997). The third perspective is on "subsidiary choice", where foreign affiliate managers are best suited to define strategies for themselves, because they are the sole owners of knowledge about the local environment within the MNC network (White and Poynter, 1984; D'Cruz, 1986; distinguishing between induced and autonomous internal development: Birkinshaw and Hood, 1997; and in a focus on 'subsidiary initiative': Birkinshaw, 1997, 1998).

Another literature that is relevant to the discussion as to under what conditions foreign affiliates draw on local knowledge sources, is the one on 'subsidiary local embeddedness'. This literature not only includes the characterisation of relationships, networks, and social capital as in Polanyi's (1957) and Granovetter's (1985, 1992) discussion of embeddedness as a mutual adaptation of the partners' perspectives, interests, and resources; it also embraces theories of networks and the innovation system literature to find that relationships are particularly important for a firm's technological development and that these have to be characterised by a high degree of embeddedness (von Hippel, 1978; Lundvall, 1988; Håkansson, 1989). Closely related is the concept of country-of-origin effects for MNCs in the international management literature. It refers to an institutional and cultural distance between the home and host economy. Far from being 'nationless' organisations, as suggested by Ohmae (1990), it is argued that even the most global MNCs still appear in many respects to be strongly rooted in their country of origin (Hu, 1992; Ruigrok and van Tulder, 1995). These effects are linked to domestic institutional and ideological structures (Pauly and Reich, 1997). Noorderhaven and Harzing (2003) argue that the country-of-origin effect manifests itself in the continued hiring of home-country nationals by the MNC, and the embeddedness of the administrative preferences of these home-country nationals in the organisational structures, procedures, and processes of the MNC. For the purpose of this analysis, the connotation of embeddedness is focused on its technological implications: the more intensely a foreign affiliate networks with local suppliers, customers, competitors, other stakeholders and institutions in the host economy in an attempt to develop knowledge and technology that is new to the foreign affiliate (hence Cohen and Levinthal, 1990), the higher the technological embeddedness of the foreign affiliate in its host economic environment (hence Lundvall, 1988).

Andersson et al. (2005) in their empirical analysis of Finland and China find that it is, importantly, the foreign investor headquarters' use of different control mechanisms that determines the extent of technological embeddedness: close monitoring by the headquarters by use of expatriate managers has a significantly negative impact on technological embeddedness of the foreign affiliate – whereas a strong emphasis from the headquarters on knowledge development by the foreign affiliate as a performance evaluation criterion has a significantly positive impact on technological embeddedness of the foreign affiliate. For the analysis presented here, the 'paradox of administration' (Thompson, 1967, p. 150) is not so very important, but rather what is deduced from Andersson et al. (2005) here is that the concept of technological embeddedness provides additional support for the proposition that the propensity of foreign affiliates to draw from the host country in terms of technology increases with affiliate autonomy.

In Andersson and Forsgren (2000), the concept of embeddedness is the intellectual starting point to explain how foreign affiliates may develop into a 'centre of excellence' within the MNC's network (see also Birkinshaw, 1997, 1998). The analysis uses 98 relationships within Swedish multinationals, and finds that the kind of management relationship between headquarters and affiliate that is most conducive to becoming a centre of excellence is not only affiliate autonomy as such. The MNC headquarters is external to the host economy network of its affiliate and hence is detached from the knowledge and technology developments generated in the foreign affiliate. For it to benefit from its technologically excellent affiliate, a business relationship of mutual 'interdependence' is needed (see also Andersson, Forsgren and Holm, 2002; Forsgren and Pedersen, 1998; Holm and Fratocchi, 1998): not only does the foreign affiliate have to be intensively embedded in the local host economy ('external embeddedness'), but it also has to be linked to the rest of the MNC network, mainly through transactions involving products, that is embodied knowledge, and disembodied technology or knowledge ('corporate embeddedness'). Note that this concept of interdependence and foreign affiliates as centres of excellence provides a more specific and empirically tested foundation to the original ideas of the main *raison d'être* of MNCs, regardless of whether this is based on transaction cost economics or resource-based theory. This, hence, suggests that strategically independent subsidiaries cooperate with other units of the MNC network and so utilise their autonomy to leverage local host economy technological assets, which will finally lead to enhanced competitive advantages of the MNC as a whole.

Another aspect of foreign investor capabilities that determines external knowledge seeking is derived from the ideas of 'subsidiary development' and 'centres of excellence' as discussed above. Yet here, a direct relationship is drawn between the foreign affiliate's technological capability and knowledge sourcing from the host economy. Frost (2001) uses the idea in which a foreign affiliate is able to assume a position of 'technical leadership' in a particular technological field (measured in patenting citation and activity), and assumes that the specialised competence that such an affiliate commands will involve some extent of technical autonomy which "allows scientists and engineers the opportunity to pursue technical paths that are more compatible with their own education, training, and experience than would otherwise be the case" (ibid., p. 105). Testing this plausibility assumption, he finds that the 'technical leadership' foreign affiliate is more likely to source knowledge and technology from its local host economy environment than from its home base. Interestingly, the results of his analysis turn out to be more robust when considering the sub-national level: local sources appear to be a particularly important basis for the foreign affiliate's own technological capability (ibid., p. 115). In the technological fields in which the foreign investor home base owns a technological advantage, home-base exploiting becomes a focal point for innovative search by the foreign affiliate rather than the local host economy (ibid., p. 106). In a further attempt to quantify these ideas of technological capability of foreign affiliates, Frost uses 'revealed technological advantage' which had previously been applied by Cantwell (1993) and Zander (1994). He finds that home base citation of affiliate patents becomes more likely in such fields where the home country owns the revealed technological advantage and vice versa (Frost, 2001, p. 117).

Furthermore, other firm-specific factors that imply firm-heterogeneity are found to determine technological embeddedness: those may include the **size** of the foreign affiliate, assuming that large firms enjoy higher legitimacy in the host country (Frost, 2001, p. 117). Typically, firm-heterogeneity would also suggest that firm **age** would play a role and positively influence the propensity and intensity of local interaction by foreign affiliates: long established foreign affiliates enjoy more intense embeddedness than newly established firms, because a positive reputation for cooperative behaviour should make access to local sources of knowledge more likely for the former whilst younger affiliates suffering from a 'liability of newness' (see for example Stinchcombe, 1965, and Venkataraman and Van de Ven, 1998, using longitudinal data spanning up to 10 years in the US). Yet, Frost (2001) is unable to find supporting evidence (ibid., pp. 188–120).

Finally, the concept of 'organisational learning' as introduced in Chapter 4 not only considers the capabilities of the foreign affiliate as the firm at the receiving end, but also considers the origins of knowledge new to the firm: the foreign parent (as in Chapter 4) and the local host economy with its innovation system (this chapter).

This already leads into the assessment of the question under what conditions foreign affiliates will draw on the local knowledge base of their host economy from the perspective of the host economy in the following chapter.

5.2.2 Capabilities of the host innovation system and spatial proximity

The objective of local knowledge/asset-seeking foreign direct investment is to tap the localised knowledge and technology base of the host economy. From the perspective of the host economy, the extent to which such foreign affiliates draw on local knowledge sources depend on mainly two distinct issues: the capability of the host economy to generate and supply relevant knowledge and technology, and, as a complementary condition, spatial proximity between the foreign affiliate and relevant actors in the host economy. Those determinants are reviewed in as general as necessary terms, but always in light of the particularities of transition economies to stay in line with the objectives of this work.

A knowledge-seeking foreign affiliate that has the capacity to adapt, implement, and take an active part in the development of host economy knowledge and technology (as discussed in the preceding sub-chapter), will still depend on the availability of a complementary knowledge and technology base in the host economy, on the ability of the host economy to allow the foreign affiliate to benefit (for example being able to communicate the knowledge and technology, Polanyi's *ability* to tell), and on the capability of the host economy to generate new knowledge in cooperation with the foreign affiliate.

The issue of capability of the host economy for foreign affiliates' technological activity can best be reviewed in the strands of literature on FDI that have evolved within the framework of the systemic approach to innovation (see most prominently Freeman, 1987; Lundvall, 1988; Nelson, 1993; Patel and Pavitt, 1994; and Metcalfe, 1995; for the concept of National[7] Innovation Systems). In the original concept of innovation systems, such a system comprises a number of different actors and institutions that play a role in developing knowledge and technology and in adapting foreign technology to function in the host economy environment. Those may include actors like suppliers, customers (users), other manufacturers (also direct competitors), science and research institutions, scientific and technical societies, policy-makers, and institutions like the labour market, the education system, financial institutions, regulatory structures, and other stake holders as well as informal institutions.

In the application of the systemic concept of innovation systems on FDI, the host economy becomes relevant as an important source of foreign affiliate technological activity. This roots in the case for cooperation and networking of the innovation literature: foreign affiliate innovation is conceptualised as driven in collaboration between the TNC and in parts by the multiple actors and institutions of the hosting economy, its host innovation system. Some of the actors have been reviewed as 'functional sources of innovation' (by von Hippel, 1988), or in the form of demand as a quantitative variable (see for example Mowery and Rosenberg, 1979 and Freeman, 1982). Institutions are assumed to set the framework conditions for networking between domestic and foreign players in the innovation systems. Those actors who expect the most attractive returns in innovation-related rents typically can be assumed to be the most important drivers or sources of technological activity of the foreign affiliate. Individual users as much as user-industries may take an active role in the innovation process for example as 'lead users', if they expect to earn a rent from being able to use an innovated product, possibly even a rent from contributing to improving a product; suppliers may develop an innovation if this would result in an increase in demand for the product that uses their inputs; other manufacturers (including direct competitors) will engage in networking if they can expect additional innovation output when using and contributing to an innovation network. For all users, networking with foreign affiliates allows them to learn and tap knowledge and technology external to the own national innovation system.

Amongst the three functional actors of users, suppliers, and other manufacturers, it is the users that have attracted particular attention (and noteworthy recently again in the focus of EU Framework Programme research project funding). The decisive conditions for their capability to play an active role for foreign affiliate technological activity is –next to the individual characteristics of 'lead users' in the functional approach (see von Hippel, 1988, Chapter 8, pp. 102–116)– the actual size of demand that users may generate: the larger the size of the market in terms of purchasing power, the larger the expected benefits from innovating according to user-specifications and user-ideas. Until today, there is little research on the availability of individuals with 'lead user'- characteristics in transition economies due to the general scarcity of research on recent user-driven innovation. But it may plausibly be assumed that amongst the population having been socialised in the socialist era, the notion that the customer (as an aggregate) eventually governs what products, services, and specifications thereof will prevail on the competitive market still remains less developed than amongst the part of the population that already grew up in a competitive economic environment. The supply with 'lead users' in transition economies will depend on the target markets of the individual product or service, and no general conclusion can be drawn (industry-heterogeneity).

For domestic suppliers and other manufacturers or competitors to be active in the technological activity of foreign affiliates, the decisive conditions include the quality of the work force, own capabilities of 'organizational learning' and 'absorptive capacities' (Cohen and Levinthal, 1989, 1990), as well as institutions. With respect to the endowment with a large and highly skilled work force in transition economies, probably the most pronounced and noteworthy observation is the divergence between formal qualification and its actual use in the occupation of the employee: common in most transition economies, the share of employees that do not use their original formal qualification in their current occupations tends to be much higher than in other countries (see for example Czarnitzki, 2005; Steffen and Stephan, 2008). This is the natural result of the profound structural changes that occurred in those countries since the beginning of systemic transition: previous working experiences were rendered irrelevant to a large extent in the newly emerging occupations, yet certificates of formal qualification remained. This is an important and largely unresolvable issue in empirical analysis involving transition economies. Another issue that becomes particularly important in transition countries with regard to the availability of skilled work force pertains to the mobility of labour between domestic and foreign owned firms: foreign affiliates tend to be financially more able and willing to pay higher wages and salaries and hence tend to attract highly skilled labour away from other firms in the host economy. This, however, weakens the host innovation system's ability to act as a competent partner for foreign affiliate development – a problem related to brain drain.

Fagerberg et al. (1994) suggest that a regions' capacity to adapt and implement new external knowledge determines the degree of attractiveness and the amount of technological spillovers it is able to draw. With regard to absorptive capacities (Cohen and Levinthal, 1989, 1990) of domestic suppliers and other manufacturers or competitors, the technology gap hypothesis or rather controversy, as well as in respect to organisational learning on behalf of all partners in the technology transfer, the reader is referred to above and in Chapter 4, in which these concepts are assessed for the foreign affiliate and with reference to CEECs where empirical insights are available. To summarise briefly: CEECs have been improving noticeably with respect to technical advance. Yet, a large share in the increasing R&D expenditures has been invested by foreign affiliates whilst domestic firms fall considerably behind.

Finally, and with respect to institutions, many formal and informal, market, and non-market institutions play a role in the strengthening of host innovation systems: mutual trust (or 'social capital') between trade partners or (technological) cooperation partners is a reflexion of diversity in norms and values between the agents from the home country and the ones in the host country. Formal institutions, like the rule of law but also discrete regulations and policies, alter the social capital between domestic and foreign players. Social capital and trust are not only of paramount importance in overseas engagement via FDI[8], they become particularly relevant where R&D, technologies, intellectual properties, and innovations are involved in trade and cooperation. The IPR regime is yet another example of a relevant institution in this respect, by regulating who will receive what share of the innovation rents (the latter issue is assessed in more detail below in Chapter 7).

In the transition economies of Central East Europe, it was the early prospect of EU membership that generated a rule of law in the process of legal transition. Accession negotiations and eventual membership with their compulsory translation of the complete EU *'acquis communautaire'* into national law served to solidify a societal and commercial environment in support of the generation of institutionally-backed trust and social capital. The adoption of the European competition law and policy as well as the regime of intellectual property rights may still show some institutional gap in the East when compared to the West, but differences have been increasingly levelling out in particular with respect to competition matters (see for example Hölscher and Stephan, 2004, 2008). Even if such formal institutions were often not sufficiently in line with informal institutions amongst the population in these countries, the achievements are certainly exceptional amongst the wider group of emerging markets.[9]

Local research institutions, publicly or privately financed, may play an active role not only incentivised by expectations of rents. Universities, and in particular where they receive a large share of public funding, typically conduct basic research out of their own will. This produces important complementary knowledge for technologically active commercial firms, and yet is often of a public character in terms of its use, even if today increasingly less so: basic research is often not *per se* easily marketable and still competition for public funds becomes increasingly competitive, hence universities become increasingly eager to cooperate with industry. Because the world of science is global, universities additionally tend to be eager to internationalise, not least by cooperating with foreign owned firms. Foreign affiliates, on the other side, may be particularly willing to cooperate with host economy universities, as they are more reliable partners, in particular, where commercially viable results give rise to legal IPR issues. Scientific research institutes, often also publicly financed, share the same advantages and may often be even closer to applied R&D. Where they are privately financed, this is typically by industry itself, and here either by a single firm (and hence the same applies here as for the above mentioned other manufacturers) or a network of firms with common research interests (then, the knowledge generated will tend to be shared in IPR pools and hence may become more of a public good character as described above for publicly financed universities). As noted above in Chapter 4, research institutions in CEECs as remnants of national academies of sciences today are less able to play an active part in the regional innovation systems (McGowan et al., 2004; Meske, 2004; von Tunzelmann, 2004; and network-failures as in von Tunzelmann et al., 2010). With regard to scientific and technical societies, those institutions are typically funded by industry, and where their activities have a local or national mandate (for example in policy-lobbying), they can be expected to be still rather underdeveloped in CEECs just because of the lower level of development of their own industries.

With a view on policy-makers as actors of the host country innovation system, it is the policies trying to attract FDI (conducted by Investment Promotion Agencies, IPAs) into the host region that play a direct and leading role for the question of attractiveness and capability of host innovation system for foreign affiliates' technological activity. Other policies like foremost innovation policies, but also industrial policies in general, and education policies, etc., are important drivers of an innovation system. In CEEC, however, they either remain less comprehensive and developed, or they form part of the mandate of IPAs.

In fact, a large share of economic policy in transition economies is tuned to the perspective of foreign investors due to the extraordinary role of FDI in those countries. Within the research on FDI-attracting policies, Birkinshaw and Hood (1997) test the role of 'host country drivers' for the 'subsidiary development process'. Those drivers include direct and indirect incentives given to FDI by the host regions' IPAs. Both turn out to be relevant incentives, even though at diminishing levels over time: at an early stage, the role of direct help and the creation of a supportive business environment for the investor turned out to be critical in some of the reviewed cases. For later stages, the development of a regional cluster turned out to be valuable in several cases, as were 'after care' programmes established in their case by the Scottish administration. It comes as no surprise that attractiveness of a region (even if generated by policies) will go hand in hand with the region's capability to play an active part in the foreign affiliate's technological development. It is not at all clear, however, whether the policies are efficient in terms of opportunity costs as long as they are effective in terms of providing relevant incentives at whatever stage. It is in particular the results of Birkinshaw and Hood (1997) referring to the development of a regional cluster in Scotland that indicate the role of policy in building the capability of (regional) host economic innovation system to the benefit of a foreign affiliate. Whilst in CEECs, the FDI-focus of such policies is gradually shifting to the needs of domestic industry, the policies are today mostly still far off being active generators of host innovation system capabilities at higher levels.

Of course other institutions in the host economies also play a role: the labour market, education system, financial institutions, regulatory structures, and other stake holders and institutions. However, their impact on the technological capability of the host economy is not as direct as the ones singled out above. Their assessment hence remain beyond the reach of this work.

Finally, the strand of literature assessing the 'contributory role' of foreign affiliates for the MNC network holds additional important insights into the capabilities of the host innovation system to provide a relevant and functioning knowledge base for foreign affiliates. Within this literature, Birkinshaw et al. (1998) perhaps somewhat counter intuitively find that the extent of local competition in the host economy, measured by own fieldwork data on a Likert scale, has a negative relationship with the contributory role of the foreign affiliate (p. 233). Whilst this stands in contrast to the pro-competitive effect in general, the authors suggest that the contributory role was particularly high for affiliates in protected niches, where the kind of contribution to the MNC network is less susceptible to upgrading through local competition (ibid., p. 235). Even if the result may have been driven by some foreign affiliates with particularly high contributory roles in such niches, their analysis does not find support for the virtuous effects of intense competition, at least not for the contributory role. If hence a foreign affiliates strives to develop a high contributory role, then the local host innovation system may be particularly capable to assist it in terms of technological development, if the foreign affiliate does not operate in an overly competitive environment. This result remains difficult to explain and it shall have to reassessed in future research.

The spatial proximity condition for foreign affiliates to source knowledge and technology in the host economy is closely related to what had been discussed in the previous Chapter 5.1 on the internationalisation of R&D. There and also here, the issue at hand is the tacitness of some of the knowledge that is sought by foreign investors (or 'sticky' as in von Hippel terms). This part of knowledge may be larger at earlier stages of the industry cycle, i.e. before standardisation in products and production processes takes place (see for example Audretsch and Feldman, 1998, 2003).

Yet, the spatial proximity condition often does not form part in FDI spillover models despite their obvious plausibility and widespread acceptance of existence of positive agglomeration effects in new economic geography (see for example Krugman, 1996) that enhance formal and informal exchange. The complex nature of technological learning and the importance of reiterated face-to-face contacts in the exchange of often tacit knowledge (see Cantwell and Iammarino, 1998, 2003) generate additional expectation as to the importance of spatial proximity for spillovers in general and the role of the host innovation system for foreign affiliate's technological activity in particular.

A number of empirical research efforts were conducted to test the importance of spatial proximity for knowledge transfer and spillovers. Not specific on FDI but on transfer and spillovers in general, Jaffe et al. (1993) analyse the geographic location of patent citations with that of the cited patents (see also Jaffe et al., 1998; Jaffe and Trajtenberg, 1998; Sjöholm, 1996). They find that the importance of knowledge sources for patents tend to increase with spatial proximity. Moreover, there is no difference between more basic and applied inventions with respect to the importance of spatial proximity. These empirical insights suggest that spatial proximity will likewise be important for transfer and spillovers between foreign affiliates and the host innovation system. This is assessed specifically by Almeida (1996) for the US semiconductor industry who establishes that technological sourcing of foreign owned firms tends to be regionally concentrated and focused on linkages between individual firms. Criscuelo et al. (2002) establish for FDI between the EU and the US in either direction that the R&D facilities' capacity to source local knowledge and technology depends on linkages between geographically co-located firms. Aitken and Harrison (1999) and Driffield et al. (2004) and Driffield (2006) conclude that intra-regional spillovers exceed the size of spillovers across regions – learning from FDI by domestic firms within and across industries tend to be local rather than national.

5.3   The data used for the analysis and testable hypotheses

Some of the propositions and expectations developed above lend themselves to an empirical analysis and to testing by use of the IWH FDI micro database of 2006–2007. The database contains information from 144 foreign affiliates in Croatia, 295 in East Germany, 110 in Poland, 220 in Romania, and 40 in Slovenia. For the current analysis, the propensity that foreign affiliates draw on the host innovation system for their own technological activity is proxied by information in the database on the foreign affiliates' level of importance attached to three sources of knowledge and technology in the host innovation system: the host scientific institutions (including universities, the remnants of state-financed academies of sciences, and commercial research institutions), suppliers, and customers. The firm-level database measures this on a Likert scale ranging from 1 for "not important" to 2 for "little importance", 3 for "important", 4 for "very important", and 5 for "extremely important".

The information contained in the database also allows to distinguish between foreign affiliates that are technologically active or not. The criterion developed from information in the database is that at minimum any one of the three technological activities have been conducted in the foreign affiliates: (i) product innovations during the years of 2002 and 2005 (in 'yes' or 'no'); (ii) the same for process innovations; (iii) annual expenditure on R&D (including external R&D) and innovation larger than 0. Innovations are defined according to the Oslo manual (OECD, 2005) and include products and processes (production or delivery methods including for example changes in techniques, equipment, and/or services) that are new to the firm, even if not necessarily new to the market.[10] Technological activity is further refined in the database by providing information about its intensity: with respect to R&D-intensities, the database contains information on the percentage share of R&D expenditure in total turnover in 2005, in total gross value added in 2005, and on the percentage share of R&D personnel in 2005 (the latter was already used in Chapter 4). Whilst R&D intensity should provide a more precise picture than pure yes/no R&D-activity, the demands on the data to generate a robust proxy over all firms in the sample are very high. Already the share of R&D employment in total employment is not easily conceptualised: not in all firms, specialised R&D personnel is employed (or work-time explicitly allocated to R&D). Very often, employees fulfil R&D tasks next to their general employment on the shop floor and during their daily work routines. The intensity of innovative activity is proxied in a comparative manner: managers of innovative foreign affiliates were asked to evaluate the individual intensities of their product and of their process innovations during the last three years in comparison to their competitors in the relevant market. The two indicators for product and process innovation intensities are ranked on a Likert scale ranging from 1 for "very low" to 2 for "below average", 3 for "average", 4 for "above average", and 5 for "very high". The database finally offers a further indicator of technological activity. The percentage "share of new or significantly improved products in the firm's total sales" may be interpreted as a very good indication of the ultimate goal of innovative activity (even if this leaves out the possibility that technological activity may be geared towards technology transfer and adoption). This indicator also adds some dynamics, as the database contains information on this indicator for the two years of 2002 and 2005 separately.

With a view on the determinants of foreign affiliates drawing on the home innovation systems for their own technological activity, a proxy is developed to allocate each foreign affiliate the level of importance of 'home-base exploiting' and 'home-base augmenting' strategies. The importance of the augmenting strategy is proxied by the "importance of the foreign affiliate as a source of technological knowledge for R&D or innovation for others", where "others" are "Headquarters of your MNE group" and "Other units or subsidiaries of your MNE group". The possible answers range from 1 for "not important", 2 for "little important", 3 for "important", 4 for "very important", and 5 for "extremely important". The proxy is defined as the average between the answers for the two (headquarters and other units of MNE group). Where the average is greater than 3.0 (i.e. on average "important" or higher), the foreign affiliate is assumed to follow a dominant augmenting strategy. The proxy for the exploiting strategy is defined as the "importance of the following sources of technological knowledge for R&D or innovation in your firm" and the possible answers have the same range as above. The two sources of technological knowledge considered for the exploiting strategy are "R&D carried out at the headquarters of your foreign investor network" and "R&D carried out by another unit of foreign investor network". Again, the average level of importance between the two is used. Where this average is greater than 3.0, the foreign affiliate is assumed to follow a dominant exploiting strategy. This way to proxy the two strategies allows not only to classify foreign affiliates as having either one of the two as dominant strategies. It also makes it possible for firms to follow both strategies at the same time (i.e. as being "important" or higher). This reflects the case made by Criscuelo et al. (2002) that some firms undertake both strategies simultaneously.

Related to this and still adding additional information, the database contains estimates on the importance of the five strategic investment motives. Because the analysis in this chapter draws from a different wave of the IWH FDI Micro Database, the list of strategic investment motives vary slightly to the ones in Chapter 3 (by use of the 2008–2009 wave): (i) accessing new markets, or increasing an existing market share (*d_market*); (ii) following key clients abroad (*d_client*); (iii) increasing efficiency across the foreign investor network (*d_effic*); (iv) accessing location-bound natural resources (*d_nat.res*); (v) location-bound knowledge, skills, and technology (*d_know*). Firms were asked to provide an indication for the level of importance for all five motives. The ranking offered in the questionnaire range from 1 for "not important" to 2 for "little important", 3 for "important", 4 for "very important", and 5 for "extremely important".

The issue of local market orientation in sales and the obvious extension of this idea to supplies, ventured here, can be quantified for analysis from information contained in the database as well. Here, foreign affiliate managers have been asked to approximate the structure of their sales and supplies distinguishing between (i) exports and imports to and from headquarters and other affiliates of the foreign investor network, (ii) to and from other foreign buyers and suppliers, (iii) to and from other domestic foreign affiliates of the foreign investor, and (iv) sales to and supplies from other domestic buyers and suppliers. The data is recorded as percentages and hence generates a continuous variable. The variables used here include the share of sales to the local market (*loc. sales*) and the share of local supplies (*loc.suppl*).

Further, information on the perceptions by foreign affiliate managers on the level of autonomy in deciding about business functions is used in this chapter again (compare the previous Chapter 4). In line with the subject of interest here being the determinants of local technology-sourcing, only the three technology-related business functions (iii) basic and applied research, (iv) product development, and (v) process engineering are considered. The indices are ranked from 1 for "no autonomy" to 4 for "full autonomy" on a discrete Likert scale.

Accounting for further firm-heterogeneity, the database finally contains information about the age (years since the foreign investor started his investment), the size of the foreign affiliate, whether the foreign affiliate is a new venture (greenfield investment), what industry (NACE two-digit) the foreign affiliate belongs to, and the host country of residence of the foreign affiliate.

What cannot be directly quantified at the firm level from the information contained in the database are indicators of the capability of the host innovation system (Fagerberg et al., 1994). Here, the analysis probes into the 'quality' of interaction of foreign affiliates with the three actors of the host innovation systems: it assumes that the host innovation capacity is high where its actors are in fact important for foreign affiliates' own technological activity. This may be a close enough proxy and has the advantage that the data has been generated from the same source. Further, the database does not contain information to test the proximity-issue. Both those 'statistical gaps' are due to confidentiality laws and the way that the field work was conducted.[11]

This leads to a set of seven testable hypotheses on the determinants of foreign affiliates drawing on the knowledge base of the host innovation system:

H(1) A foreign affiliate following an augmenting strategy is more likely to rely on the host innovation system as a source of knowledge for its own technological activity.

H(2) A foreign affiliate following an exploiting strategy is less likely to rely on the host innovation system as a source of knowledge for its own technological activity.

H(3) The higher the share of local supplies of a foreign subsidiary, the more likely it is to source technological knowledge from local suppliers.

H(4) The higher the share of local sales of a foreign affiliate, the more likely it is to source technological knowledge from local customers.

(H5) The higher the autonomy of a foreign affiliate, the more likely it is to source technological knowledge from the host innovation system.

H(6) The longer the foreign affiliate is established in the host economy, the more likely it is to source technological knowledge from the host innovation system.

H(7) The larger the foreign affiliate is in terms of employment, the more likely it is to source technological knowledge from the host innovation system.

The analysis below includes all CEECs in the database by use of information from the IWH FDI micro database.

5.4 The importance of the host innovation system for foreign affiliates' own technological activity

The objective of the following analysis is to use the IWH FDI micro database and where available additional locational statistics to empirically test the hypotheses above. This helps to answer the second research question of this chapter (RQ5_2) on the identification of empirically significant determinants of a positive role played by the host innovation systems in CEECs for technological activity of their local foreign affiliates. The empirical tests are preceded by some descriptive representation of the most relevant aspects at hand for which the database contains information from the sample foreign affiliates.

5.4.1 Descriptive statistics on own technological activity of foreign affiliates

The chapter is concerned with the extent of own technological activity of foreign affiliates in CEE. The multiplicity of technological activity-indicators, using information about foreign affiliates' R&D and innovation activities, does not only increase the robustness of information, it is conceptually necessary: R&D and innovation may not always follow patterns of a linear relationship. Next to its usual interpretation as a condition to generate innovations in a concept of linear causation, R&D-activity may follow objectives and give rise to effects quite independent of innovation. Cohen and Levinthal (1989) teach that R&D not only stimulates innovation (not even necessarily as a precondition, as some innovations are generated by mere chance), but R&D also develops the ability to identify, assimilate, and exploit external knowledge (which is a precondition for technology transfer and spillovers at the receiving end). In a test of the relationship between R&D expenditure (both in absolute values and as per cent of turnover) and product and process innovations by way of pairwise correlation analyses, not one coefficient turned out to be of a substantial magnitude (the highest significant coefficient is 16.6 between R&D as a share of turnover and product innovation intensity).

The simultaneous use of information on employment in R&D and expenditures for R&D reflects the condition that for significant technological activity, foreign affiliates should be expected to invest into R&D using financial resources and at the same time to have explicitly nominated personnel for R&D. After all, firms may record R&D expenditure even without R&D personnel, for example if R&D is contracted outside the firm (in the field work, expenditures for external R&D services were explicitly included). The following descriptions of each of the two indicators separately allows to identify external R&D, where the difference between expenditure and personnel is large[12]. Foreign affiliates located in East Germany, Poland, and Slovenia appear to have the highest shares of R&D-active foreign owned firms both in regard to employment and expenditure (see Figure 5.1). In particular, the shares are above 50 per cent of the total number of firms interrogated in the field work that led up to the database. Foreign affiliates located in Croatia and Romania are much less likely to be research-active.

Information about the intensity of R&D activity is of course much more telling than information that only distinguishes between activity or no

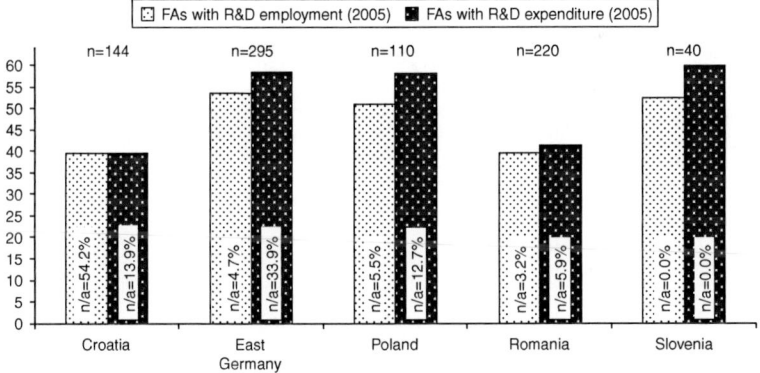

Figure 5.1 R&D activity of foreign affiliates across countries: share of research-active foreign affiliates

Note: The shares represent those firms that do have R&D employment or R&D expenditure respectively. n/a is the share of firms that have not provided information with respect to R&D employment or R&D expenditure. n is the total number of foreign affiliates per host economy.

Source: IWH FDI Micro Database 2006–2007.

activity. But, it is obvious that this information was not always immediately at hand immediately during interviews, hence the extremely high standard deviations of country-specific averages (see Figure 5.2). Even if, therefore, the averages presented are not very robust, they do show large differences between host countries, and in particular all three indicators alike show the same tendencies: share of R&D employment in total employment, share of R&D expenditure in turnover, and in gross value added.[13] This does in fact suggest that the results are reliable and sound.

Amongst East German and Croatian foreign affiliates, the share of work efforts devoted to R&D are highest with nearly 9 per cent, whilst this figure is only around 2–3 per cent for Polish and Romanian affiliates. In Slovenia, foreign affiliates on average spend around 4–5 per cent of their work efforts on R&D. This ranking in three groups remains largely unchanged when looking at the share of R&D expenditure in gross value added (calculated as the value of total sales corrected by the share of intermediate inputs/supplies). In the case of R&D expenditure in per cent of turnover, East German foreign affiliates on average move away from the level of Croatian firms and closer to the level of Slovenian affiliates, which may well be an effect of larger firm-sizes in East Germany as compared to Croatian or Slovenian foreign affiliates.

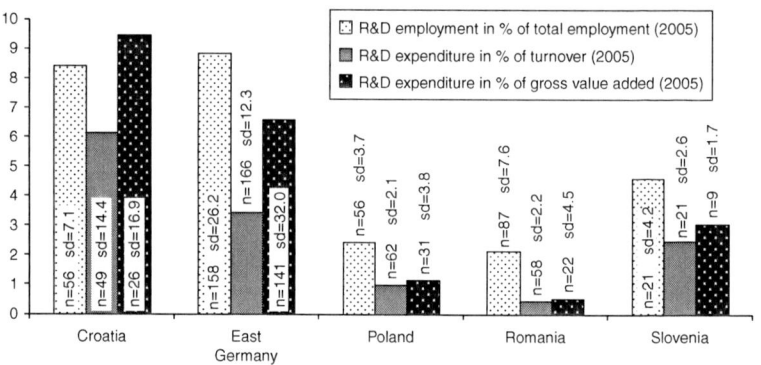

Figure 5.2   R&D intensity of foreign affiliates across countries

Note: Averages of firm-specific shares of R&D employment or expenditure are calculated as aggregated means over the country-groups of FAs that provided information on R&D activity (i.e. with individual weights for FAs and excluding n/a). n is the number of R&D-active FAs per host economy. sd denotes standard deviation. In the cases of East Germany and Romania, one statistical outlier each had to be removed.

Source: IWH FDI Micro Database 2006–2007.

In sum, the R&D-related data suggest that East German, Polish and Slovenian foreign affiliates are more likely to be active in terms of R&D. This is mirrored by R&D intensity in East Germany and Slovenia, whilst in Poland, the large share of research-active foreign affiliates is not paralleled by an equally high intensity of R&D. Croatian foreign affiliates, on the other hand, are less likely to be research-active and yet appear to be much better endowed with R&D personnel and to spend more on R&D than the Polish or Slovenian affiliates.

Contrasting Croatia and Poland, the results suggest that R&D is more scattered between many foreign affiliates whereas in Croatia, R&D is more concentrated amongst a few very R&D active firms. Romanian foreign affiliates have the lowest indicators for R&D activity across the board.

Some of this very general picture is confirmed by the data on the innovativeness of foreign affiliates, measured by the share of innovators in total country-specific sub-panels, and by estimates of foreign affiliates' managers about the intensities of product and process innovation in comparison to competitors in relevant markets.

Figure 5.3 presents the share of foreign affiliates in each of the host countries of the database that has produced innovations according to the above conceptualisation. It is again East Germany, Poland, and Slovenia that have the highest shares of innovative foreign affiliates, whereas Croatia and Romania fall behind.

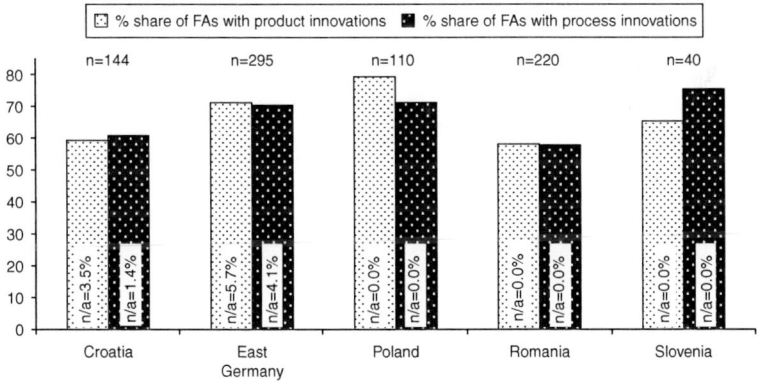

Figure 5.3 Innovative activity of foreign affiliates across countries: share of innovative firms

Note: The shares presented here are the firms that provided the information that they have undertaken innovations. n/a denotes the share of firms that have not provided information with respect to innovations. n is the total number of foreign affiliates per host economy.

Source: IWH FDI Micro Database 2006–2007.

In respect to innovation intensity, however, the picture changes somewhat (see Figures 5.4 and 5.5): when using the share of foreign affiliates that consider their innovative activity as "above average" as the cut-off point *vis-à-vis* everything at average levels or below, then the perception of foreign affiliate managers on their comparative innovation intensity of product and process innovations is highest amongst Polish affiliates, followed by East German affiliates. Croatian, Romanian, and Slovenian foreign affiliates falling further behind. Whilst this ranking applies to both product and process innovations, Slovenian foreign affiliates fare even worse with respect to process innovations (see Figure 5.5) compared to product innovations (see Figure 5.4).

The result for Poland is clearly driven by extraordinarily high shares of firms estimating an "above-average" intensity of product and process innovation, whilst the category of "very high" comparative intensity has been selected seldom by Polish managers. The somewhat better result for Romania for process innovations is driven by an extraordinarily high share for the "very high" category of intensity – this may in fact be a country-specific difference in perception.

In the end, however, this suggests that East German and Polish foreign affiliates are not only both more likely to be innovators but also have high innovation intensities. Croatian and Romanian foreign affiliates

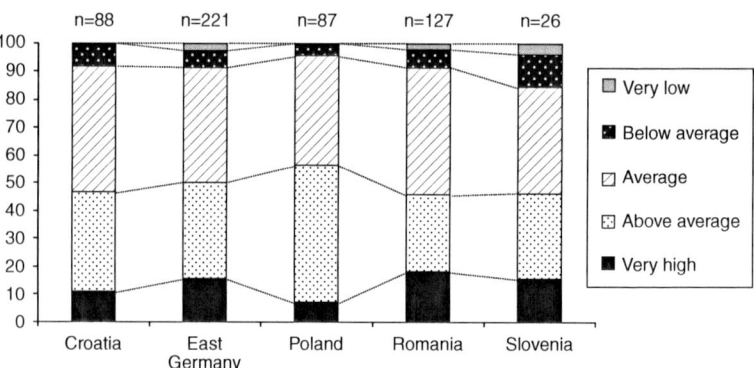

Figure 5.4   Innovative activity of foreign affiliates across countries: intensity of *product* innovation

Note: Cumulative bars represent the shares of *product* innovation-active firms that compare their intensity of *product* innovations with those of their competitors in the relevant markets to be very low, below average, average, above average, and very high. n/a denotes the share of *product* innovation-active firms that, however, have not provided information with respect to their comparative innovation intensity.

Source: IWH FDI Micro Database 2006–2007.

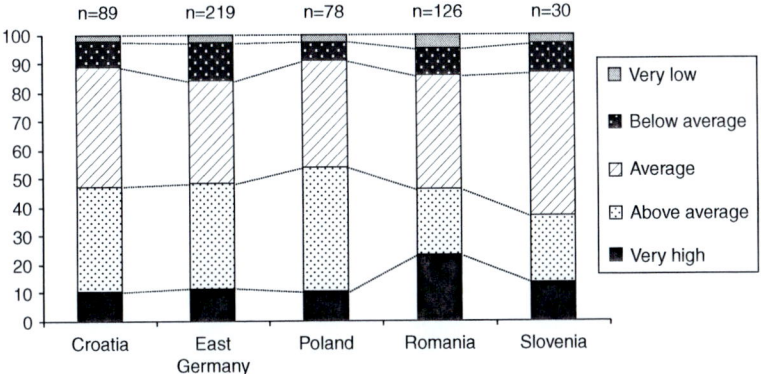

Figure 5.5 Innovative activity of foreign affiliates across countries: intensity of *process* innovation

Note: Cumulative bars represent the shares of *process* innovation-active firms just as above. n/a denotes the share of *process* innovation-active firms that, however, have not provided information with respect to their comparative innovation intensity.

Source: IWH FDI Micro Database 2006–2007.

are neither highly likely to be innovative nor have high average innovation intensities. Foreign affiliates of Slovenia, however, produce a less consistent message: the high share of innovative firms does not translate into an equally high intensity of innovation.

It is the latter result that stands in contrast to the picture portrayed by the data on R&D: whilst Slovenian foreign affiliates on average appear to be rather research-active in all indicators, their innovativeness does not reflect this. Moreover, their comparatively high propensities to be active in terms of R&D and innovation does not correspond to equally high R&D and innovation intensities. In the cases of East Germany with comparatively high average throughout and Romania with lower averages over all indicators, a corresponding picture between R&D and innovation is drawn. In Croatia, the high R&D intensity is probably concentrated amongst only few firms, and R&D efforts translate into a rather smaller share of innovative firms with rather lower levels of innovation intensity. Amongst Polish foreign affiliates, the high probability to be active in R&D does not translate into an equally high R&D intensity, whereas the high probability to be innovative does in fact translate into equally high innovation intensities. This puzzle can be further tested by use of the data on the "share of new or significantly improved products in the firm's total sales", presented in Figure 5.6.

The most obvious result in the descriptive representation of the indicator on the result of innovative activity is that the comparatively high levels of technological activity from above also translate into a comparatively high share of sales by innovative products. Over time, this has even become more pronounced. In the case of Romania, the comparatively low levels of technological activity correspond to a low share of innovative products in total sales in 2002, but by 2005, this has increased 4-fold, the data does not provide any more information to disentangle this. Amongst Polish foreign affiliates, the share of innovative products in total sales mirrors the picture for innovation intensity, even if R&D intensity was low: the many research-active firms are able to generate sales based on new products, even if the amount of R&D going into the generation of new products is low. Whilst this suggests a high level of efficiency in the use of scarce R&D resources, it does raise a question mark with respect to the sustainability of technical advance from Polish forces. Croatia's concentration in the structure between technologically active and not active foreign affiliates, paired with low levels of innovativeness and innovation intensity equally results in low average shares of innovative products in domestic sales. Slovenia's rather inconclusive picture of technological activity and intensity at comparatively high levels finally results in

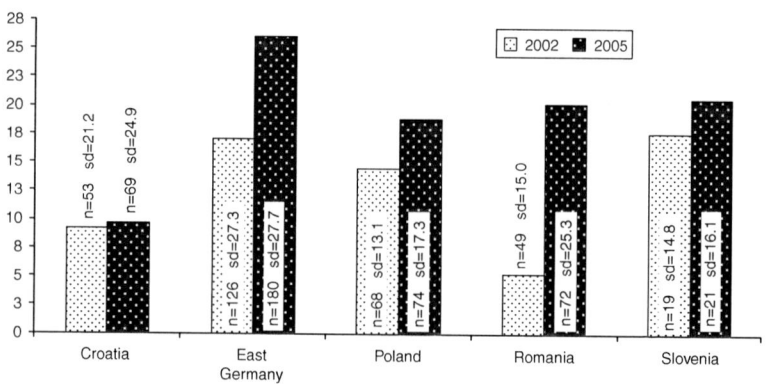

Figure 5.6　Aggregated share of sales (in %) attributable to new or significantly improved products across countries

Note: The aggregate shares presented here are the sum of sales attributable to new or significantly improved products per the sum of firms' sales, using only firms that have provided the information for either figure (n). sd denotes standard deviation.

Source: IWH FDI Micro Database 2006–2007.

high shares of new products in total sales which supports a rather high technology interpretation of this host country. Across all countries apart from Croatia, the shares of sales of new products has increased markedly between 2002 and 2005. Again important to notice is that not only standard deviations are in some cases very high indeed, but also the number of FIEs that make up these aggregated shares vary between 2002 and 2005.

The information presented on the R&D and innovation activities and outcomes of foreign affiliates provides a comprehensive picture of own technological activity of FDI affiliates within host countries. This clearly brushes over any heterogeneity within host countries, hence the meticulous reporting of the number of cases and standard deviations. Furthermore, the quality of this information depends on the representativeness of the sample of foreign affiliates in the transition countries of the IWH FDI database. Because, however, there is no comparable data available on these issues that cover FDI projects in these countries, this represents the best information available. Following on from this country-comparison, the analysis proceeds to enquire into the determinants of technological activity with particular reference to the involvement of the host innovation system.

5.4.2 Descriptive statistics on the determinants of host knowledge sourcing

The first factor that is expected to determine local host economy knowledge sourcing for foreign affiliate technological activity pertains to March's (1991) distinction between exploitation and exploration strategies. Figure 5.7 depicts the share of foreign affiliates in the database that either predominantly follow an exploiting strategy (i.e. the proxy for exploiting is greater than 3.0 and that for the augmenting is lower than 3.0), or a dominant augmenting strategy (i.e. proxy for augmenting greater than 3.0 and exploiting lower than 3.0), or both strategies as important, very important, or extremely important (i.e. both proxies greater than 3.0), or finally none of the strategies (i.e. both proxies lower than 3.0).

The results show that FDI projects in the database in fact often envisage both strategies as important (supporting the case of Criscuelo et al., 2002): the shares of foreign affiliates attaching importance to both strategies at the same time evolve around 45 per cent amongst Romanian foreign affiliates, around 35 per cent for Croatian, around 30 per cent amongst East German and Slovenian foreign affiliates, and nearly 20 per cent for Polish foreign affiliates.

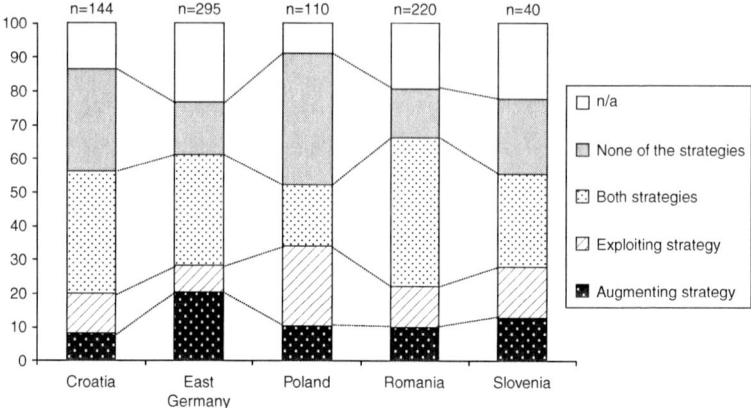

Figure 5.7 Importance of exploiting and augmenting strategies across countries

Note: The information is provided in shares of firms per host economy that value the exploiting or the augmenting strategies as important, very important, or extremely important. The shares not adding up to 100% represent those firms that have not provided the information. n is the number of firms that have provided information on the questions used to proxy the strategies.

Source: IWH FDI Micro Database 2006–2007.

Assuming that foreign affiliates following March's exploration (or home-base augmenting) strategy are more likely to draw on local knowledge sources than firms with an exploitation strategy, the descriptive statistic suggests that interaction with the host innovation system is most intense in East Germany, followed by a large margin in Slovenia, Poland, Romania, and Croatia. Interestingly, amongst the Polish FDI projects, a large share of around 40 per cent (Croatian around 30 per cent) do not attach any particular importance to either augmenting or exploiting strategies – little knowledge transfer involved in strategies is to be expected here.

Related to the distinction between those two strategies are the investment motives, and yet they do add additional information. Table 5.1 lists averages and the corresponding standard deviations of averages of importance attached by foreign affiliates to strategic motives in groups of host economies. The results are reassuringly close to the results from the interrogation of investment motives in the database of 2008–2009 (even if the motives are slightly less fine-graded in the earlier version of the database here, see Chapter 3): from the perspective of the foreign affiliate, access to new markets or the increase of an existing market share turns out be the dominant strategic motive in all economies but

Table 5.1 Strategic investment motives of foreign affiliates, by country

	Market	Follow client	Efficiency	Natural resources	Knowledge, skills, techn.
Croatia	**3.5** (1.4)	2.1 (1.2)	**3.5** (1.3)	2.2 (1.4)	3.1 (1.2)
East Germany	**3.3** (1.3)	1.9 (1.3)	2.8 (1.3)	1.7 (1.2)	3.0 (1.3)
Poland	**3.8** (1.1)	2.3 (1.3)	3.3 (1.1)	2.5 (1.4)	2.5 (1.0)
Romania	**3.4** (1.4)	2.6 (1.4)	3.3 (1.3)	2.5 (1.5)	3.0 (1.4)
Slovenia	3.0 (1.6)	2.0 (1.2)	**3.2** (1.4)	1.5 (1.0)	2.7 (1.2)
Total	3.4 (1.4)	2.2 (1.3)	3.2 (1.3)	2.1 (1.4)	2.9 (1.3)

Notes: Unweighted averages over the country-groups of foreign affiliates. The numbers in parenthesis are standard deviations. The number of firms are 106 for Croatia, 257 for East Germany, 107 for Poland, 213 for Romania, 40 for Slovenia. The highest levels for motives within each host country are highlighted in bold.

Source: IWH FDI Micro Database 2006–2007.

in Slovenia in 2005. Here, the motive to increase efficiency across the foreign owner network is slightly higher than the host market motive. The efficiency-related strategic motive turns out to be the second most important one for Poland, Croatia (by a very small margin), and Romania. These are incidentally also the countries with the lowest wages amongst the countries assessed here.

Of particular interest for this chapter is of course the strategic motive to tap the local host economy knowledge base in terms of existing knowledge, skills of personnel, and available technology. This motive appears to be particularly important for East German foreign affiliates (second only to the market motive), less so for Slovenia, Croatia, and Romania (all 3rd rank). For foreign investors into Polish affiliates with their large home markets and low labour costs, access to local knowledge and, skills and technology appears to be the least important one on average. Over all foreign affiliates in the country-samples, this motive is third after market access and increasing efficiency.[14] Standard deviations appear to be low enough to warrant robust interpretation of results.

Cantwell and Mudambi's (2005) local market orientation in sales and the obvious extension of this idea to supplies, follows the assumption that host country customers will gain in importance as sources of knowledge and technology for the foreign affiliate if it has intense trade relations to such local customers. The same applies, vice versa, to suppliers: the foreign affiliate is more likely to draw on local suppliers in the host innovation system economy as sources of knowledge and technology for its own technological activity if a large share of its supplies originates from local suppliers.

Figure 5.8 depicts the average shares of local supplies and local sales of foreign affiliates in the database. East German foreign affiliates appear to be sourcing and supplying most intensively from the host region, with rates for supplies and sales both above 50 per cent. Whilst this is also the case for Poland, the share of local supplies are here somewhat lower than in East Germany. The shares of Croatia, Romania, and Slovenia are all distinctly lower. In the case of Slovenia, it is also noteworthy that domestic sales assume a much lower share than domestic supplies – a reflection of the small market in terms of purchasing power and yet a more able domestic industrial base.[15]

The literature on local embeddedness raises the issue of corporate control of foreign affiliates. The assumption to be drawn from this literature is that foreign affiliates with high levels of autonomy from parent companies are more likely to source knowledge from their host innovation systems (see for example Andersson et al., 2005). Figure 5.9 presents the averages of autonomies between the three technology-oriented business functions (basic and applied research, product development, and process engineering) for each host economy.

The results suggest, perhaps unexpectedly, that foreign affiliates in Romania appear to be the most autonomous in the three

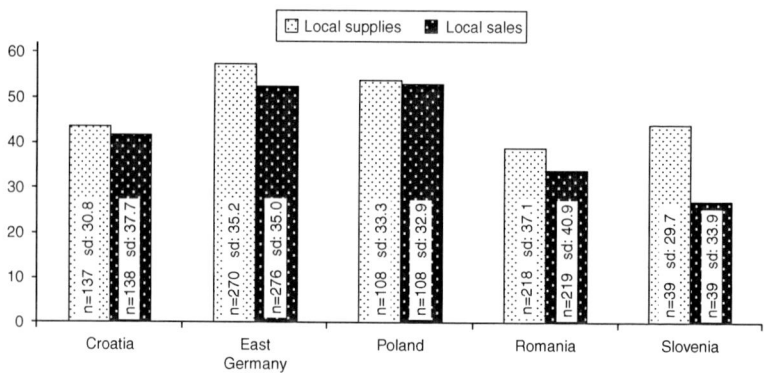

Figure 5.8 Average shares of local supplies and sales of foreign affiliates across countries

Note: The averages are computed from firm-individual shares and are presented in percentages. n is the number of firms that have provided information on the relevant questions. sd denotes standard deviation.

Source: IWH FDI Micro Database 2006–2007.

technology-related business functions. This is mainly due to comparatively much higher autonomy levels in 'basic and applied research'. Incidentally the average level of autonomy of Romanian foreign affiliates is also higher than in the other host economies if all seven business functions are considered (see above in Chapter 4). This suggests a high degree of independence in deciding about own technological activity, even if R&D intensity is lowest in comparison to the other countries (see above).

Slovenian, Polish, and Croatian foreign affiliates command much lower levels of autonomy, whilst East German foreign affiliates are somewhere in the middle. The multivariate analysis below tests whether the high level of autonomy for Romanian foreign affiliates and the low level for, in particular, Polish foreign affiliates still have the kind of repercussion on local host economy embeddedness that the literature suggests.

Also taken from the local embeddedness literature, the expectation is that the age of the foreign affiliate and its size will have an impact on the intensity of drawing on the host innovation system as a source of knowledge and technology (Frost, 2001). The assumption is that technology-sourcing from the host innovation system increases with age (or the time since the foreign investment was made) and the size of the foreign affiliate (the 'liability of newness' case). The descriptive statistics presented in Figure 5.10 show that Croatia has on average the most recent FDI projects (years since the foreign investment was made and as of 2005), which may not come as a surprise due to the war involving

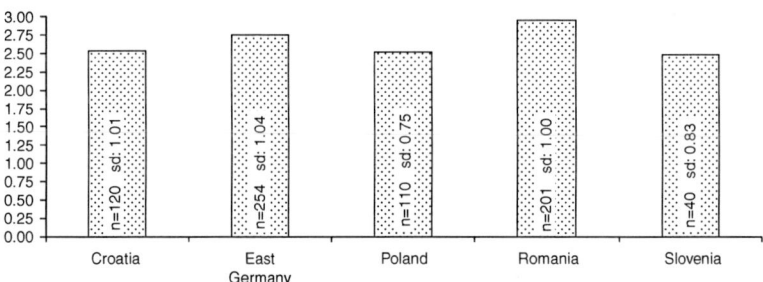

Figure 5.9 Average levels of autonomy in three technology-related business functions of foreign affiliates across countries

Note: Autonomy levels are defined as 4 for maximum autonomy and 1 for minimum autonomy. n is the number of firms that have provided information on all three business functions. sd denotes standard deviation.

Source: IWH FDI Micro Database 2006–2007.

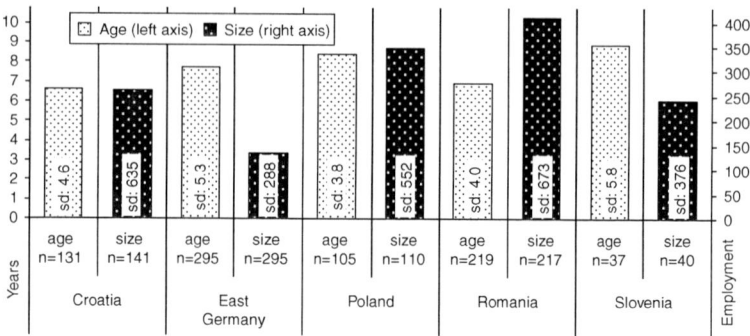

Figure 5.10   Age and size of foreign affiliates across countries

Note: Age of foreign affiliates is defined as years since the foreign investment was made as of 2005. Size of foreign affiliates is measured in total employment. n is the number of firms that have provided information. sd denotes standard deviation.

Source: IWH FDI Micro Database 2006–2007.

the former Yugoslav regions. Slovenia had been able to break away from this conflict relatively peacefully and very early on, and consequently has on average older foreign affiliates. It is however astonishing that between all host countries assessed here, the Slovenian average turns out to be the highest. Even when discounting one foreign affiliate that originates in 1972, the average is still as high as in East Germany, where most foreign investments were undertaken in the first part of the 1990s, but are now below the average of Poland. Romania can be termed a transition latecomer, having only joined the EU in 2007. Accordingly, the country's FDI projects are of more recent date.

In terms of size, the differences are even more pronounced: Romania has the largest FDI projects with an average employment of around 411, East Germany the smallest with around 135, and the other countries have between 250 and 350 employees.[16] East Germany may stand out here due to the particular privatisation method, which involved the splitting up of (former) state-owned enterprises into commercially more viable units during the times of the *Treuhandanstalt*; a large share of the 'restructured' smaller units were subsequently bought by West German investors. In the other countries, large-scale privatisation programmes involving foreign capital for the selling of larger firms dominated the scene for a long time.

Finally, the question of foreign affiliates drawing on the host innovation system is exemplified here by the three most prominent actors in the host innovation systems, the local scientific institutions, local suppliers, and local customers. Scientific institutions include not only universities with their more basic research facilities and the remnants of state-financed academies of sciences, but also commercial research institutions financed by industry.

Figure 5.11 shows that the average levels of importance attached to the three actors of the host innovation system do not vary very much between the actors in the cases of East Germany and Poland, a rather consistent picture. In East Germany, the scientific institutions are rated somewhat higher than suppliers and also a bit higher than customers, which is related to the quality of the scientific institutions here. It is not surprising that customers would assume an important role for foreign affiliate technological activity due to the size of the market in terms of purchasing power. In Poland, the size of the market is mainly driven by the number of consumers, not by the comparatively lower purchasing power (if GDP per capita in purchasing power parities is used as an indicator). Consumers rank the highest amongst the three actors of the Polish innovation system, a testament to the comparatively high importance attached to the market-seeking strategic motive of FDI projects in this country. In Romania, scientific institutions are valued by foreign affiliate managers much less highly as sources of knowledge and technology for their own technological activities than are local suppliers and customers. In fact, between all five FDI hosts, Romanian scientific

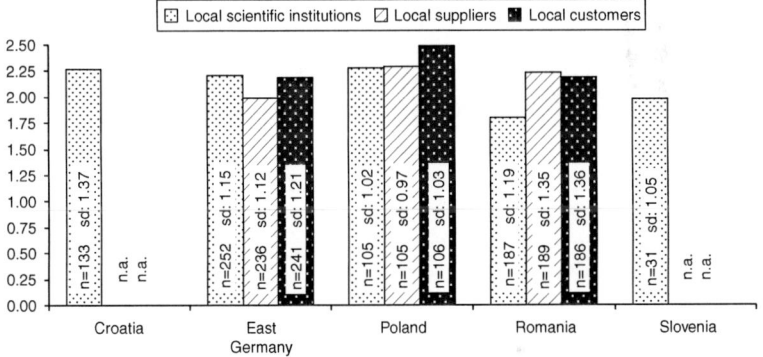

Figure 5.11 Averages of levels of importance of host innovation system actors for foreign affiliates' technological activity

Note: Unweighted averages over all firms in one country. n is the number of firms that have provided information; for customers and suppliers, data is not available in the Slovenian and Croatian subsamples. sd denotes standard deviation.

Source: IWH FDI Micro Database 2006–2007.

institutions rank lowest, with a level of 1.79, outdone by Slovenian scientific institutions (1.97) and the three remaining hosts. The levels of Croatia, East Germany, and Poland are all very similar, at around 2.20 and 2.27. Unfortunately, the data does not contain any information on local suppliers and customers in the Croatian and Slovenian samples.

Whilst the country-specific pattern comparing Romania and the other countries or regions is largely in line with our expectations (Romania being a 'transition latecomer'), the similar levels of importance attached to Croatian, East German, and Polish scientific institutions does come as a surprise, considering the dissimilar states of transformation and development of these institutions in those countries. In particular, the expectation was that the East German scientific institutions should have been valued much more highly by foreign affiliate managers (see the explanations on the state of transformation of the former academies of sciences by McGowan et al., 2004; Meske, 2004, von Tunzelmann, 2004; von Tunzelmann et al., 2010; mentioned several times above).

These country-specific descriptions already suggest that the host economy alone cannot explain differences in the intensities of use of host innovation systems for foreign affiliates' own technological activity. Rather, heterogeneity at the micro-level dominates. This can be accounted for by a firm-level multivariate regression analysis of determinants of technological interaction of foreign affiliates with local scientific institutions. The econometric analysis tests the hypotheses generated in the conceptual part above, and for which the database offers relevant proxies.

5.4.3   The method of empirical analysis

The empirical analysis is based on the theories and concepts described in the review above: a set of seven determinants is used to test the hypotheses specified above. To account for firm-heterogeneity, the analysis additionally controls for investment motives, greenfield investments vs acquisitions, the industry of the foreign affiliate, and the host country. This contributes to answering the second research question (RQ5_2) of this chapter.

The issue of foreign affiliates drawing on the host innovation system forms the dependent variable in the regression analysis. In line with the availability of data, the host innovation system is represented in the analysis by scientific institutions, suppliers and customers, where those reside in the host country (or in the case of East Germany: host region of the former GDR).

Because the dependent variables of importance of host innovation system actors form discrete variables on a five-point rating scale, an ordered probit method is applied in the regression analysis (this follows the suggestions of Wooldridge, 2002; and Greene, 2003). In particular, the dependent variable ranks different outcomes where distances between outcomes are not necessarily identical or known. In its theoretical form, the model reads:

$$y^* = x'\beta + \varepsilon$$

where y^* is the unobserved endogenous variable, β is the parameter vector for x', and ε is the error term (assumed with a mean equal to zero and variance equal to unity). The real y is unobserved because the answers are given only in discrete values that come closest to the real y of the person interviewed: in statistical terms, the answers of interviewees are hence defined as:

$$y = 0 \; if \; y^* \leq 0$$
$$y = 1 \; if \; 0 < y^* \leq \mu_1$$
...
$$y = J \; if \; \mu_{J-1} \leq y^*$$

The μ_{J-1} cut-off points are the unknown parameters to be estimated with β. The likelihood function takes the form:

$$p(y = 0|x) = \varphi(-x'\beta)$$
$$p(y = 1|x) = \varphi(\mu_1 - x'\beta) - \varphi(-x'\beta)$$
...
$$p(y = J|x) = 1 - \varphi(\mu_{J-1} - X'\beta)$$

Greene (2003) argues that it is a sufficient assumption that the distribution is known and continuous, just as for all maximum likelihood estimations. Possible heteroskedasticity is corrected for by using White robust standard errors (heteroskedasticity-robust standard errors) in all regression models, and accounts for the possibility that the error terms have not the same variance across all observations.

The model is specified with a parameter vector including:

HIS = augm; expl; loc.sales; loc.suppl; autono3; age; size; d_market; d_client; d_effic; d_nat.res; d_know; d_greenf; d_indus; d_host

with *HIS* denoting the importance attached to any one of the three actors in the host innovation system; *augm* and *expl* being the two

Marchian strategies of 'exploiting' and 'augmenting'; *loc.sales* the share of sales to the local market; *loc.suppl* the share of supplies procured on the host markets; *autono3* for the degree of autonomy in the three technology-related business functions; *age* in years since the foreign investment was made and as of 2005; *size* in employment numbers; *d_market* as the dummy variable controlling for the investment motive of access to markets that assumes a value of 1 where the foreign affiliate manager values this motive as "important", "very important" or "extremely important" and 0 otherwise; *d_client* controlling for following foreign key clients; *d_effic* for increasing efficiency across the foreign investor network; *d_nat.res* for accessing location-bound natural resources; and *d_know* for accessing location-bound knowledge, skills, technology.[17] *d_greenf* controls for FDI projects involving the establishment of new firms; *d_indus* controls for the 14 NACE two-digit single main industrial occupation of the foreign affiliate; *d_host* for the five of the countries of location of the foreign affiliate in Central East Europe (Croatia, East Germany, Poland, Romania, Slovenia). Because local knowledge sourcing is only relevant where foreign affiliates are in fact technologically active, the analysis only considers foreign affiliates that have undertaken any product innovations over the last three years.[18]

Before regression models are tested, multicollinearity between determinants is checked pairwise, correlating all independent variables. Table 5.2 shows that none of the significant correlations are in fact at levels that would raise problems for the regression analysis. The highest and significant correlation coefficients result between the exploiting strategy and the augmenting strategy with a positive sign and .44 (suggesting that both strategies appear to be followed simultaneously by a number of firms), and between the exploiting strategy and the proxy for technological autonomy with a negative sign and -.41 (which is not a surprising result). Local sales and local supplies likewise have a comparatively high and statistically significant rank correlation coefficient with .33, which supports the assumption that foreign affiliates' integration into their host economies pertains to a simultaneous interaction with customers and suppliers.

The quality of the ordered probit models is assessed by use of the Pseudo-R^2 (that is interpreted as being bound between 0 and 1) and furthermore by use of Wald tests (under the assumptions of consistency and asymptotic normality, see White, 1982).

The concept underlying this regression model is not derived from a general theory but rather from the above review of international business, international management, and innovation bodies of literature,

Table 5.2 Pairwise rank correlation between all determinants

	augm	expl	loc.sales	loc.suppl	autono3	age	size	d_market	d_client	d_effic	d_nat.res	d_know
expl	.44*											
loc.sales	-.08*	-.02										
loc.suppl	-.09*	-.10	.33*									
autono3	-.12*	-.41*	.09*	.22*								
age	-.01	.07	.07	.03	-.10*							
size	.03	.08*	-.22*	-.15*	-.04	.07						
d_market	.14*	.23*	.25*	.08*	-.07	-.08*	-.04					
d_client	.10*	.10*	-.08*	-.10*	-.10*	-.01	.02	.26*				
d_effic	.20*	.22*	-.22*	.10*	-.20*	.01	.10*	.18*	.16*			
d_nat.res	.06	.09*	.06	.09*	.02	.06	.12*	.18*	.10*	.15*		
d_know	.16*	.03	-.07	.8*	.06	-.04*	-.09*	.06	.03	.18*	.21*	
d_greenf	.12*	.01	-.04	-.08*	-.15*	.13*	-.21*	-.05	.09*	.01	-.02	-.08*

Note: All pairwise correlations were conducted as Spearman's rank correlations using case-wise deletion, where observations are ignored if any of the variables in the list of variables are missing: obs=489. Significant correlations (at a level of .1) are marked with an asterisk.

Source: IWH FDI Micro Database 2006–2007.

as well as empirically tested plausibility assumptions. Hence, the model conceptualisation contains some extent of eclecticism. It remains uncertain which determinants are necessary and sufficient to explain the dependent variable, and hence several model specifications are tested, adding credibility to the assessment of robustness of results: model 1 includes all determinants that are assumed to have an impact (and where the database offers data), including country dummies and industry dummies. Model 2 is estimated without controlling for strategic motives, model 3 without industry dummies, model 4 without country dummies, and finally model 5 is tested without either country, or industry, or strategic motive dummies. Also burdened with some uncertainty is the direction of causality between the determinants of importance attached to host scientific institutions and the technological activity of the foreign affiliate itself. Without being able to fully account for this endogeneity problem by way of suitable instrumental variables, the analysis uses only foreign affiliates that have undertaken product innovations over the last three years.

5.4.4 Estimation results

Table 5.3 presents the results of the ordered probit estimations for the importance attached to host scientific institutions (the determinants of local sales and local supplies are not considered here, because they are assumed not to play a role for the importance of scientific institutions as such). This is exemplarily tested in a number of alternative model specifications to determine the best performing specification for subsequent analyses of all three actors in the host innovation system.

All models tested are significant overall, Wald tests confirm that all independent variables are significantly distinct from 0 (Prob > χ^2 and Wald χ^2). All models have very low levels of pseudo R^2 which may, however, still be acceptable: ordered probit estimations typically have very low pseudo R^2 and the data in the database contains considerable heterogeneity due to the fact that these are all subjective assessments of foreign affiliate managers. Model 1 has the highest pseudo R^2 with about .08, which is a typical result arising from the inclusion of all relevant variables and dummies. As expected, the model with the lowest number of determinants tested also results in the lowest pseudo R^2, of approximately .06. The number of observations is lower than the total population in the database: of the 566 technologically active foreign affiliates, the models are able to use only 500 firms due to missing values. Across all model specifications, the resulting significant and insignificant determinants remain largely the same; the only changes appear for the

Table 5.3 Determinants of host scientific institutions assuming an important role for foreign affiliates' own technological activity – the analysis of CEECs

	Model 1	Model 2	Model 3	Model 4	Model 5
augment	.10* (.058)	.14* (.012)	.10* (.067)	.13* (.015)	.15* (.005)
exploit	.30* (.000)	.32* (.000)	.30* (.000)	.29* (.000)	.31* (.000)
autonomy	.01* (.000)	.01* (.000)	.01* (.000)	.01* (.000)	.01* (.000)
age	−.01 (.182)	−.01 (.212)	−.01 (.224)	−.01 (.262)	−.01 (.301)
size	.00 (.646)	.00 (.782)	.00 (.610)	−.00 (.594)	−.00 (.577)
d_market	.00 (.973)	–	.01 (.896)	.04 (.738)	–
d_client	.20 (.143)	–	.18 (.160)	.10 (.435)	–
d_effic	.15 (.158)	–	.14 (.179)	.12 (.268)	–
d_nat.res	.26* (.072)	–	.22 (.107)	.17 (.214)	–
d_know	.22* (.063)	–	.21* (.060)	.20* (.086)	–
d_greenf	−.17 (.113)	−.18* (.095)	−.17 (.109)	−.23* (.036)	−.24* (.024)
d_indus	yes	yes	–	yes	–
d_host	yes	yes	yes	–	–
n	500	500	500	500	500
Wald χ^2	121.38	88.71	98.70	107.24	62.91
Prob > χ^2	0.000	0.000	0.000	0.000	0.000
Pseudo R^2	0.088	0.076	0.078	0.075	0.057

Note: Coefficients with * and in bold are significant at the .1 level. Figures in brackets denote heteroskedasticity-robust P > |z|.

Source: IWH FDI Micro Database 2006–2007.

greenfield investment dummy that loses significance in models 1 and 3 (albeit a very near miss in both cases) and the dummy for the motive to access natural resources, which is only significant in model 1. Changes in signs never occur.

Comparing all five alternative models, it appears that model 1 is performing the best, with the highest Pseudo R^2 (even if the greenfield variable loses significance). This suggests that country dummies and industry dummies do in fact play a role, and controlling for them is important in the following sets of analyses. When testing the determinants of other host innovation actors assuming an important role for foreign affiliates' own technological activity, the model specification of model 1 is used. Table 5.4 presents the results for suppliers and customers, and it reproduces the results for scientific institutions for easier comparability. In the models with suppliers and customers as dependent variables, the corresponding determinant of sales or supplies on the host market respectively is included (data on the importance of suppliers and customers are not available in the Slovenian and Croatian subsamples, reducing the number of observations in the tests to levels around 370–380).

Of the seven topical determinants (*augment*; *exploit*; *local sales*; *local supplies*; *autonomy*; *age*; *size*), the two strategies of 'home-base exploiting' and 'home-base augmenting', local sales, local supplies, and the autonomy in the three business functions signal quite consistent results. Apart from the 'home-base augmenting' strategy in the model

Table 5.4 Determinants of host innovation system actors assuming an important role for foreign affiliates' own technological activity – the analysis of CEECs

	Local scientific institutions	Local suppliers	Local customers
augment	.10* (.058)	.12* (.072)	.07 (.281)
exploit	.30* (.000)	.31* (.000)	.18* (.012)
loc.sales	–	–	.01* (.000)
loc.suppl	–	.01* (.000)	–
autonomy	.01* (.000)	.01* (.000)	.01* (.017)
age	–.01 (.182)	.01 (.464)	–.01 (.481)
size	.00 (.646)	–.00* (.014)	–.00 (.405)
d_market	.00 (.973)	–.18 (.193)	–.15 (.267)
d_client	.20 (.143)	.03 (.838)	.21 (.168)
d_effic	.15 (.158)	.28* (.024)	.26* (.031)
d_nat.res	.26* (.072)	.25 (.153)	.11 (.501)
d_know	.22* (.063)	–.03 (.840)	.26* (.051)
d_greenf	–.17 (.113)	–.27* (.036)	–.22* (.087)
d_indus	yes	yes	yes
d_host	yes	yes	yes
n	500	372	377
Wald χ^2	121.38	102.30	103.03
Prob > χ^2	0.000	0.000	0.000
Pseudo R^2	0.088	0.091	0.087

Note: Coefficients with * and in bold are significant at the .1 level. Figures in brackets denote heteroskedasticity-robust P > |z|.

Source: IWH FDI Micro Database 2006–2007.

for customers, coefficients are significant and consistently with a positive sign. The results for age and size, however, remain unclear: the test cannot significantly hold that coefficients are distinct from 0, and the tests produce no support for a statistically significant influence of either the size or the length of establishment of a foreign affiliate in the host economy. Only size turns out to be significant but negative in the model for suppliers.

With regard to the above hypotheses, the tests find significant support for H(1) holding that with increasing importance of the augmenting strategy, foreign affiliates also find it increasingly important to source knowledge and technology from actors in the host innovation system (this at least with respect to scientific institutions and suppliers). This implies that those foreign affiliates actively search for new knowledge beyond the established knowledge base of the foreign investor, as was also found by Frost (2001) in his analysis of US greenfield subsidiaries between 1980 and 1990. It also suggests that foreign parents that mandated their foreign affiliates with a home-base augmenting approach to innovation and technology development strive to enhance the firm-specific advantages of the whole corporation by their own technological activity. This is in line with the findings of Cantwell and Mudambi, 2005, on R&D intensity of UK foreign affiliates of non-UK multinational enterprises.

Hypothesis H(2) assumes that a foreign affiliate that is following an exploiting strategy is less likely to rely on the host innovation system as a source of technological knowledge. Alas, the results of the empirical tests clearly refute this by finding significant support for the converse: the statistically significant coefficients consistently have a positive sign and for all three actors: scientific institutions, suppliers, and customers. This unexpected result may, however, be explained rather easily, as it reinforces the argument of the Criscuelo et al. (2002) analysis of patent citation data on FDI between the US and the EU: foreign affiliates typically follow both augmenting and exploiting strategies for different technologies at the same time. In other words, sourcing knowledge from the parent in one technological field often goes hand in hand with external technological interaction and local knowledge sourcing in another field.

Hypotheses H(3) and H(4) hold that because local market orientation could entail customisation of product and process technologies to the needs, preferences, possibilities and particularities of the local market, foreign affiliates with a high share of local supplies and local sales can be expected to source knowledge and technology from host suppliers and customers respectively. The higher the level of local demand, the higher the incentive to undertake process improvements as well as to differentiate output (as found out by Cantwell and Mudambi, 2005, on UK-based FDI). Foreign affiliates may also benefit from technological spillovers through backward vertical linkages, because the benefits from vertical technology transfer are assumed to go both ways (as Hoekman and Smarzinska, 2006, deduce from their summary of empirical literature, and as Giroud, 2007, finds for multinationals in Malaysia and Vietnam). Both hypotheses are in fact supported by the analysis here on FDI in CEECs: a high share of local procurement significantly increases the probability that foreign affiliates rely on local suppliers as a source of knowledge for their own technological activity. Likewise, a high share of local sales tends to increase the importance of customers as actors in the host innovation system.

In regard to the corporate governance issue, the empirical tests suggest that the probability of local sourcing of foreign affiliates increases with the autonomy of the foreign affiliate from its parents in terms of business functions. This supports H(5) and is in line with the results of Roth and Morrison (1992) and Birkinshaw and Morrison (1995), and in more general terms the concepts of subsidiary development, roles and embeddedness (see the many sources (referenced above) having evolved around scholars such as Bartlett, Birkinshaw, Ghoshal, Hood, and others).

The empirical tests above do not produce significant results for the two reputation-related hypotheses on age H(6) (based on Frost, 2001) and size H(7) (going with Stinchcombe, 1965, and Venkataraman and Van de Ven, 1998) of the foreign affiliate. Even so, it remains important to control for those firm-specific particularities to account for firm-heterogeneity. In an attempt to account for further firm-specific heterogeneity, the empirical tests additionally control for strategic investment motives by use of five dummies (one for each investment motive) for greenfield investments for the industry the foreign affiliate is mainly engaged in, and the analyses finally also control for the host economy. The coefficient for the dummy for efficiency-seeking investment motives turns out to be positive and significant for the importance of both suppliers and customers. Local knowledge-seeking foreign affiliates are significantly more likely to source knowledge from scientific institutions and customers, but not, however, from suppliers, which puts a tentative question mark behind suppliers as capable actors in the host innovation systems of the countries in the analysis. The dummy for greenfield investment turns out to be negative and significant in two of the three models (and still close to significance in the third), suggesting that possibly the 'liability of newness'-hypothesis, using age, that the analyses were unable to support still plays a role and is caught by the greenfield dummy.

So far, the data available in the IWH FDI Micro Database does not include regional determinants to test for possible effects arising from spatial proximity and from varying capabilities of the regional host innovation system. In a related analysis by Günther et al. (2008b), such an extension of the analysis was carried out, but only for the East German section of the database. The first additional variable used in this analysis pertains to regional endowment with human capital, proxied by HRSTO employment shares. The second variable proxies the quality of the scientific infrastructure of the host region. The third variable represents the regional knowledge stock, and the fourth and final additional variable measures agglomeration effects. The results are very interesting but often remain inconclusive and are in need of further qualification. Still, the levels of pseudo R^2 are markedly higher than for the analyses of the whole sample presented above, a sign of country heterogeneity that was not caught by the country dummies in the previous analysis. Most importantly, the results suggest firstly that the endowment with human capital has a statistically significant effect for the likelihood to source technology and knowledge from local customers, but not from local scientific organisations or local suppliers. The lack of significant results is possibly rooted in the definition of the proxy, which restricts human capital to the same sector the foreign affiliate operates in and excludes vertically related sectors. Second, the role of scientific infrastructure is tested positively for the cases of scientific institutions and suppliers – not for customers. Arguably, R&D expenditure by higher education institutions will predominantly affect the capability of scientific institutions and suppliers. The capability of customers to act as an important source of knowledge and technology may be assumed to be less affected by quality of the scientific infrastructure. Third, the analysis establishes that the regional knowledge stock is a determinant of the capability of the regional host innovation system, and is significant and positive for customers, as expected. The results for scientific institutions and suppliers, however, are surprising: here, the results suggest that a large knowledge stock reduces the probability of sourcing knowledge and technology from local scientific institutions and suppliers. Again, the answer may lie in the design of the proxy: where patent intensity is high (in the relevant region and the relevant industry at NACE two-digit level), technological competition may be assumed to be likewise high, and technological interaction with firms and institutions outside the

MNC network may involve too many dangers. This may in particular be relevant with respect to scientific institutions (where the sharing of IPR rights is an issue) and with respect to suppliers, who will sometimes supply not only the foreign affiliate alone but possibly also other firms in the same industry – competitors. This effect may not exist for market-oriented foreign affiliates, where customers are end users and not very likely to become competitors. Finally, the results indicate a negative relationship between local knowledge sourcing and agglomeration advantages: in regions with lower agglomeration densities (in terms of employment within a particular sector compared to other regions), foreign affiliates are more likely to source technological knowledge from the customers and scientific institutions. Lacking plausible explanations for this result beyond the above suggested competition effect, the relevance of agglomeration effects remains unresolved in this analysis: the proxy may possibly compound intra- and inter-sectoral effects, or the results represent agglomeration disadvantages due to congestion (which seems rather implausible in the case of East Germany).

In sum, the East German region is a particular case of a transition region with important factors setting this region apart from other the CEECs analysed earlier, yet the comparable results largely remain the same. The regional indicators, however, produce mixed results that call for further research effort to test the plausibility explanations suggested.

5.5 Summary of main results

The analysis in Chapter 5 to a large extent supports the empirical results found for other host regions around the world, independently of whether those constitute developed, emerging, or developing host countries or regions. This shows not only that FDI in CEECs behaves according to the rules and experiences found elsewhere, despite the important particularities of that region; it also suggests that the data and methods applied in this chapter are valid and robust.

The global trend of intensifying internationalisation of technological activity within TNCs (or its increasingly decentralised organisation) progressively includes emerging markets, and here also CEECs as technology-generating locations. Even if this may not always include core technologies of the TNC or commercially valuable R&D, it still serves to improve the potentials of a positive technological role of FDI in CEECs.

With respect to the first research question of this chapter (RQ5_1), the reviews identified several important conditions of foreign affiliates drawing on local knowledge sources. Those include in particular foreign investor strategies, whereby the 'home-base augmenting', or 'competence augmenting' strategy is associated with much more local host economy knowledge sourcing than 'home-base exploiting' or 'competence exploiting' behaviour. The former reiterates Cantwell's theory of technological accumulation, where the foreign affiliate actively contributes to developing its host region(s). Likewise, a local market orientation of foreign affiliates is associated with more host innovation system interaction than export orientation (or even OPT-type) of the FDI project.

The concept of subsidiary development holds that autonomy is an important determinant of interaction with the host innovation system, just like the quality of the host economic environment, and the extent to which the foreign affiliate is the sole owner of knowledge about this host economy environment. The concept of technological embeddedness is likewise concerned with the management relationship between foreign affiliate and its investor, but holds that a form of mutual interdependence is most conducive to the foreign affiliate developing into centres of excellence. Here, size and age of the foreign affiliate also play an important role in determining local knowledge sourcing.

A second aspect of RQ(5_1) pertains to the capability of the host innovation system. Here, the potentials of the market and users, lead users and user-industries play an important role. In CEECs, this factor is becoming increasingly relevant due to the growth of their economies and resulting growth in purchasing power. With respect to the supply of human capital in CEECs, the gulf between formal qualifications and current occupations is only partly bridged by working experience, and the brain drain from domestic firms to foreign affiliates is an important issue in CEECs. Furthermore, the state of development of the national research institutions and industry associations is still rather low in CEECs, reminiscent of the former socialist era. Policies in CEECs that would potentially be able to help host innovation systems to excel in terms of capabilities are today still often focused on foreign affiliates, unable to generate host innovation system capabilities at higher levels. Other institutions, such as for example mutual trust (social capital) and IPR also are found to play an important role for local knowledge sourcing. In terms of spatial proximity, little research is available, yet proximity has turned out to be important where included in empirical analysis.

The analysis of technological activity (R&D and innovation) of foreign affiliates in CEECs suggests that whilst East German, Polish, and Slovenian foreign affiliates are more likely to be active in terms of R&D, it is Croatian and East German foreign owned firms that appear to be better endowed with R&D personnel and spend more on R&D than do Polish or Slovenian affiliates. Romanian foreign affiliates have the lowest indicators for R&D activity across the board. With regard to product and process innovation, Poland signals a high level of efficiency (in the linear concept where R&D leads to innovations) with low inputs in terms of R&D and a high share of sales based on innovations (as output).[19] The comparatively high levels of East German technological activity translate into a comparatively high share of sales in innovative products; the correspondence of effort and output draws a solid picture of East German technological development.

In the case of Slovenia, the statistics of technological activity (in terms of propensity to engage in R&D and innovation, and the intensity of R&D efforts and innovative output) remain somewhat inconclusive. In the end, however, Slovenian foreign affiliates are able to generate a large share of turnover from new products, which suggests a positive interpretation of this host country as a technological location. Croatian and Romanian foreign affiliates, however, rank somewhat lower in terms of technological activities and the resulting commercial results. All countries exhibit growing shares of new products in total sales between 2002 and 2005; only in the case of Croatia is this negligible.

To a large extent, these results correspond with the general wisdom about the technological capabilities of the CEECs. Hence the results here are quite in line with expectations: East Germany is typically seen as the technologically most advanced host economy in the group of transition economies. This is a positive result of German re-unification, featuring early access to the EU and substantial financial and institutional transfers from West Germany. Slovenia has the highest level of economic (and likewise technological) development, Poland is attractive for foreign investors due to its large market and low wages, and Romania has only recently begun to catch up. Croatia, economically burdened by the setback of the war, trails behind.

With respect to RQ5_2 on the significant determinants of a positive role played by the host innovation systems in CEECs for technological activity of their local foreign affiliates, the results of the econometric analysis suggests that country dummies and industry dummies play a role, and controlling for them is important. Yet, pseudo R^2 are low, suggesting a large extent of heterogeneity still not captured in the determinants included in the analysis (a comparable analysis of East Germany results in much higher pseudo R^2 which shows that there is still a large degree of heterogeneity between host countries that is not caught by the country dummies).

Further, the analysis finds significant support for the hypothesis that foreign affiliates in CEECs with a dominant home-base augmenting strategy actively search for new knowledge beyond the established knowledge base of the foreign investor; and from the perspective of the foreign investor, that those foreign affiliates contribute to enhancing the firm-specific advantages of the whole corporation by local knowledge sourcing.

The analysis also finds support for Criscuelo et al. (2002) that knowledge sourcing from the parent in one technological field often goes hand in hand with external technological interaction and local knowledge sourcing in another field: both home-base augmenting and exploiting strategies are significantly involved with local knowledge sourcing. Furthermore, local knowledge sourcing is driven by local market orientation in terms of supplies and sales to and from the local host economy: the importance of local suppliers as sources of knowledge and technology rises with the share of procurement from local sources (vertical upstream), and the importance of local customers rises with the share of sales to local customers (von Hippel's user argument, and quite in line with the assumptions about OPT).

With respect to the autonomy issue in corporate governance between foreign investor and affiliate in CEECs, the results support the general view in the international business literature (concepts of subsidiary development, roles and embeddedness) that local knowledge and technology-sourcing increases with rising autonomy. Finally, the analysis suggests that greenfield investments appear to suffer from 'liability of newness'-sclerosis in terms of their ability to source local knowledge and technology. Beyond the characterisation of foreign affiliates as greenfield investments, newness is not a significant determinant of local knowledge sourcing in terms of age or size.

The analysis remained thus far unable to include regional determinants to test for possible effects arising from spatial proximity and from varying capabilities of the regional host innovation system. The related analysis briefly described in Günther et al. (2008b) suggests, however, that the endowment of the host region with human capital in fact increases the propensity of foreign affiliates to source knowledge and technology from local customers (but not from local scientific institutions or local suppliers). The results are just the other way round for the role of the local endowment with scientific infrastructure; the results pertaining to the size and quality of the local knowledge stock indicate that technological competition is high in this region, and that this reduces the desirability of technological interaction with local firms and institutions outside the MNC network. This is easily conceivable with respect to scientific institutions (where the sharing of IPR rights are an issue) and local suppliers (who may also supply competitors). The same interpretation involving the dangers of losing ownership advantages to (potential) competitors is suggested for the role of agglomeration advantages on the propensity of local knowledge sourcing, which also turns out to be negative in this related analysis.

6
Foreign Affiliates as Knowledge Source for Host Regional Innovation Systems in CEE

The analysis in this chapter complements the analysis in the previous one: the technological-embeddedness issue introduced in the previous chapter on the determinants of sourcing knowledge and technology from the host innovation system is not uni-directional: embeddedness can also involve the flow of knowledge and technology from the foreign affiliate to other actors in the host innovation system. This is within the most widely researched area of the developmental role of FDI and is best known as spillover literature.

The most basic observation in this issue is that the diffusion of knowledge and technology is costly (Teece, 1976), in particular, where knowledge and technology assume tacit characteristics. Costs are of relevance to the foreign investor, where it transfers its ownership advantage to foreign affiliates and where it receives new knowledge from its foreign affiliates (internal technology transfer), for foreign affiliates, where it causes knowledge and technology to dissipate to the host economy (whether intended or not) and where it receives knowledge from the host economy and other firms in the foreign investor's network or the host economy. Finally, costs arise for host economy firms, where they both send or receive knowledge from foreign affiliates (external transfer and spillovers). For the issue of the developmental effect of FDI, it is the direction of technology diffusion from foreign investor network to foreign affiliate to the host economy that is of particular importance (that is the pipeline model, for example Caves, 1974; Globerman, 1979; Haskel et al., 2002; and Keller and Yeaple, 2003). The issue of costs involved in the diffusion of knowledge and technology that arise in the host economy as the receiving end can be conceptualised as the absorptive capacity of the host economy firms and institutions (Cohen and Levinthal, 1989, 1990). And still, it is not only costs that determine the extent of knowledge and technology transferred from FDI projects to the host economy through various chapters, and it is those determinants that are the subject of analysis in this chapter. It is hence the potentially benevolent effects of FDI via technology and knowledge diffusion that are the focus of this chapter, not any other, even though arguable, important effects (on growth, employment, etc.), because of the technology focus on the entire work. The research question of this chapter is hence:

RQ(6) What are the empirically significant determinants of a positive role of inward FDI for the technological activity of the host innovation systems in CEECs?

This research question is assessed again by way of an econometric analysis testing a set of hypotheses on the role of determinants and channels of diffusion of knowledge and technology from foreign affiliates to the host economy. This is preceded by a very brief overview of what the available literature on CEECs holds in terms of the existence of spillover effects, their sign and their magnitude.

The connotation of 'spillovers' in the economic literature can be slightly different from the meaning of the word in some of the international business literature: in the former, spillovers involve market failure, the inability of the source of knowledge generating spillovers to fully internalise the value of the benefits that accrue to the recipient(s). In the international business literature, however, this externality aspect does not often play a (conceptually explicit) role. Where it does play a role, the focus is not on the benign effects of spillovers (contributing to a positive developmental role for FDI), but rather on the dangers involved with the unwanted dissipation of 'ownership advantages' in the Dunning sense. A distinction is often made in the international business literature between technology transfer and spillovers, not separated –as would be consistent with economic theory– according to the externality criterion, but in the sense that technology transfer denotes the exchange of knowledge and technology between the parent investor and the foreign affiliate. Spillovers then describe the diffusion of knowledge and technology between the foreign affiliate and the host economy, mostly domestic firms. Because the analysis here is not concerned with distinguishing between unwanted and intended knowledge transfer or spillovers, the connotation of knowledge spillovers is used here to denote MNC-external diffusion of knowledge, regardless of whether it is of a 'transfer' or 'externality' type.

6.1 The current state of the art in research

Many of the concepts and theories discussed above in the chapter on the role of the host innovation system for the foreign affiliates' own technological activity are also relevant to the other direction of the flow of knowledge and technology. The following is hence a very brief sketch of where the insights and plausibilities summarised in the previous chapter either apply here as well, or have to assume a different perspective.

The foreign investor strategy of augmentation (March, 1991; Frost, 2001; Cantwell and Mudambi, 2005; Kuemmerle, 1997) is not only more likely to grant the host innovation system an important role for technological activities of the foreign affiliate, but it is also more likely to involve sizeable diffusion of knowledge and technology from the foreign affiliate to the host economy: where FDI projects seek to use the local knowledge base to augment the ownership advantages of the foreign investor's network, the use of knowledge and technology from the host economy will effect some reciprocal flow from the foreign-owned firm to the host economy. Only here, the diffusion of knowledge and technology is not as explicitly enshrined in the concept of the two strategies as the opposite direction of flows. With respect to the exploitation strategy, the implications for the issue of spillovers may not be clearcut (and, as it turned out in the previous chapter, a significant and positive determinant of the opposite direction of flow of knowledge and technology): foreign affiliates with a dominant exploitation strategy (where FDI projects concentrate on using the internal MNC stock of knowledge and technology and only add little technological development to align their products and services to the particularities of the host economy) may still carry significant new knowledge and technology from which the host economy may benefit. This is particularly so where exploiting FDI projects work with technology of a higher standard or carry knowledge so far unavailable in the host economy. Where such FDI projects adjust their own activities to the environment of the local host economy, new knowledge and technology are developed explicitly and are of high relevance to the host economy.

The distinction between local-market orientation vs export orientation (Cantwell and Mudambi, 2005) may turn out to be rather different for the analysis of a direction of flow of knowledge and technology from the foreign affiliate to local firms. Whilst a local-market orientation of FDI projects was expected to be associated with more local-knowledge sourcing than export orientation of FDI projects, the expectations for the direction of knowledge flows assessed here may be quite different: foreign affiliates with an export orientation may be assumed to be of higher technological value to local firms than foreign affiliates with a local-market orientation; being oriented towards exporting may help to 'demonstrate' to local firms the kind of knowhow necessary to conquer external markets. CEECs as host economies are also emerging markets with a lower purchasing power for technologically sophisticated products, so that a local-market orientation may involve less technological upgrading of products than if western markets are the prime targets of FDI projects hosted by this region. This can be assumed to be particularly pronounced where a local-market orientation targets final-end consumers, and possibly less where customers use the products and services of the foreign affiliate as inputs into their own production (here, more technology may be involved, in particular where customers form part of other MNC networks).

The implications of the concepts of 'subsidiary development' (literature around Bartlett, Birkinshaw, Ghoshal, Hood, and Morrison), or 'centre of excellence' (literature around Andersson, Forsgren, Pedersen), and 'subsidiary local embeddedness' (literature around Lundvall, Håkansson) on the interaction with the local economy may be expected to carry the same signs for either direction of knowledge flows, whereby upgraded or upgrading foreign affiliates that are technologically embedded may not only draw more intensively from the host economy but may also generate more knowledge diffusion from the foreign affiliate to the host economy.

With respect to the capabilities of the host innovation system, the role of demand as a driver of technical change (for example, von Hippel) may not play an equally important role for the diffusion of knowledge to domestic firms (in particular where products and services are sold to end consumers), as it would for the opposite direction of knowledge flows. Much more so, however, do absorptive capacities (Cohen and Levinthal, 1989, 1990) play an important role for spillovers: knowledge and technology that may diffuse from foreign affiliates to domestic firms will only be useful if properly understood and applied and adjusted at the receiving end.

Finally, any exchange of knowledge and technology between two organisations necessitates direct human interaction and hence spatial proximity, regardless of direction of flow.

At the most general level, the considerable body of empirical literature on spillovers remains somewhat ambiguous with respect to the identification of spillovers, their direction, and their size in individual cases. The large number of reviews (both narrative or in the form of more or less formalised meta-studies) is both a sign and a result of the many contradictory empirical results. The difficulties involved in measuring spillovers do not come as a surprise, not least due to their character of not being remunerated (at least in the strict sense, see the above discussion of the connotation of 'spillovers'), or even due to problems associated with the data and empirical methods applied in researching spillovers. But that science should still remain unable to agree on general conventional wisdom despite the many research and review efforts requires a good explanation, and that – according to the opinion of the author of this work – is best coined by Lipsey and Sjöholm (2005): "the main lesson might be that the search for universal relationships is futile" (ibid., p. 40) which of course is reminiscent of heterogeneity between FDI projects, host countries, time etc.

6.1.1 Spillover channels

In the knowledge and technology spillover literature (for some examples of the many reviews available, see Clark et al., 2011; Blomström and Kokko, 1998; Erdogan, 2011; Görg and Strobl, 2001; Görg and Greenaway, 2004; Jindra, 2006; Meyer and Sinani, 2009; Reddy and Zhao, 1990, Wooster and Diebel, 2010), a number of channels are identified and empirically tested: the first channel pertains to the mobility of labour. Here, former employees of foreign affiliates that may have received training and that have experienced the knowledge and technology of the foreign affiliate can move to domestic firms (see for example Fosfuri, Motta and Ronde, 2001; Glass and Saggi, 2002a; Meyer, 2003 in an application of this for vertical spillovers).[1] A further channel is the 'demonstration effect' whereby domestic firms become more efficient by imitating production technologies as well as management and marketing techniques of foreign firms (see for example Das, 1987; Wang and Blomström, 1992). A 'competitive effect' is assumed where the entry of MNC-foreign affiliates increase the intensity of competition in relevant host markets (see for example Dunning, 1993; Caves, 1974; Markusen and Venables, 1999; Wang and Blomström, 1992). Again, a counter-effect is conceivable if the entry of a new (foreign) competitor forces a reduction in the efficiency of scales of production, or where domestic firms are even driven out of the market. The reasoning behind this as a technological effect of inward FDI can best be described in terms of the typical (short-term) firm-specific cost curves at the micro-level that assume positive economies of scale (see Figure 6.1).

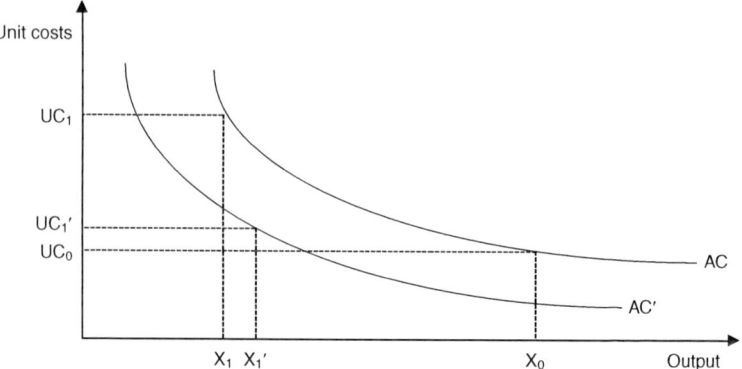

Figure 6.1   Average (short-term) cost curves of domestic firms in a situation of entry of foreign-owned competitors

Source: Adapted from Torlak (2004).

If entry of a foreign-owned firm competes away market shares of a domestic firm, the resulting fall in output of the domestic firm ($X_0 \rightrightarrows X_1$) implies higher unit costs ($UC_0 \rightrightarrows UC_1$): an initial competitive disadvantage (assumed to root in some sort of Dunning's ownership advantage) results in a further deterioration of competitiveness, with lower scales (X_1) and higher unit costs (UC_1). However this reasoning not only assumes significant economies of scale but also portrays a zero-sum game, where foreign-owned and domestic firms compete for a given market share. Moreover, this conceptual representation remains static (at most static-comparative, if the firm-specific average cost curve AC shifts downwards to AC', with the corresponding X_1' and UC_1'), inasmuch as it assumes constant technologies (or insufficient improvements of technology): the domestic firm is unable to improve its technology (sufficiently) to become efficient and competitive. If the latter is the case, then the competition effect is in fact benign, inasmuch as it drives out inefficient economic activity (see the 'pro-competitive effect', and the Schumpeterian concept of 'creative destruction'). The negative interpretation of the competitive effect is at times rather misconceived in the literature: the issue at hand is not so much a possibly detrimental effect of inward FDI as an implication of a lack of international competitiveness of domestic enterprises. In terms of pure welfare, this is a benign effect. And yet of course in terms of a national view on the economy, there may in fact be a case against dominant foreign ownership, but this is an issue of a different field of science, not of the technological role of inward FDI. Clearly, this channel is more relevant for horizontal spillovers.

Knowledge spillovers may also occur via foreign trade (exports in particular). Domestic firms can benefit from foreign affiliates through a reduction in entry costs into foreign markets by way of imitation or cooperation with foreign affiliates. Imports may also be a trade-related channel, if foreign direct investors import capital goods with embedded technology that domestic firms may buy or use, and from the importation of specialised inputs into production involving domestic firms (see for example Aitken et al., 1997; Rhee, 1990).

Finally, and closer to the point of interest in this analysis, knowledge and technology may spill over via vertical or horizontal linkages between foreign affiliates and domestic firms (see for example Lall, 1980; Rodriguez-Clare, 1996; Markusen and Venables, 1999; Pack and Saggi, 2001). Within vertical linkages, backward linkages diffuse knowledge to suppliers where the foreign affiliate as customer requires an upgrade in product quality by supplying technical assistance, worker training or machinery, or in services by providing managerial or organisational support (Blomström and Kokko, 2003, Smarzynska Javorcik, 2004a). Downstream linkages may generate technological benefits for customers, particularly for semi-finished products where the supplying foreign affiliates transfer embodied technology or technical assistance accompanying products. Horizontal or intra-industry spillovers will typically not be made intentionally, because the foreign affiliate will typically tend to prevent rather than promote the diffusion of its knowledge advantage to (potential) host economy competitors.

6.1.2 Empirical analysis of horizontal spillovers in CEECs

Horizontal spillovers have received a large share of attention from scientists empirically testing existence, magnitude, and determinants of spillovers in CEECs. Such spillovers are assumed to emerge where domestic firms in the same industry gain access to the ownership advantage of the foreign affiliate (whether intended or not, which is an issue of the IPR regime). Also, the demonstration effect may be particularly relevant for the horizontal kind of spillovers. The large body of empirical literature on CEECs produces diverging results (for a recent review and meta-analysis of horizontal spillovers in developing countries, see Wooster and Diebel, 2010). The typical empirical studies use a measure of productivity (usually apparent labour productivity or TFP via growth accounting) in a production function framework and either test over time with new FDI as an event, or cross-sectionally between industries/countries with vs without

FDI. The presence of FDI is usually measured in terms of employment rather than invested capital due to the correspondence of the former to personal interaction, assumed to be vital for spillovers.[2]

A group of authors comprising Jože P. Damijan, Mark Knell, Boris Majcen, and Matija Rojec extensively researched technology transfer and spillovers in CEECs. In a study of eight CEECs[3] over the period 1994–1998 using firm-level data, Damijan et al. (2003a) compare direct technology transfer through FDI, intra-industry knowledge spillovers from FDI, firms' own R&D accumulation, and R&D spillovers through trade for TFP growth of local firms. They establish that technology is transferred primarily via internal transfer between foreign investor and affiliate whilst intra-industry spillovers are not identified to a significant extent. In a further study (Damijan et al., 2003b), now involving all ten CEECs (having added Latvia and Lithuania) and over 90,000 firms over the period 1995–1999, do find horizontal spillovers, and yet the impact on firms' productivity by way of internal technology transfer turns out to be "by some factor 50 larger than the impact of backward linkages of FDI and by factor 500 larger than the impact of horizontal spillovers" (p. 17) (see also Chapter 4). This implies that backward link-ages are larger than horizontal linkages by a factor of 100. The analysis further establishes that absorptive capacities and productivity gaps play a very important role in the process of technology transfer and spillovers. Two of the authors (in Damijan and Knell, 2003a) focus on Estonia and Slovenia by use of input–output tables and firm-level data on nearly 1500 firms, and again remain rather critical with respect to the existence of significant spillovers. Their analysis reiterates the domi-nance of internal technology transfer and foreign trade as sources of foreign knowledge and technology, rather than spillovers.

Another research group around Holger Görg also remains rather critical with respect to spillovers in CEECs. In a literature review, with David Greenaway, of empirical research on spillovers in various regions in developed, developing, and transition economies, they conclude: "robust empirical support for positive spillovers is at best mixed" (Görg and Greenaway, 2004, p. 171). See also Görg and Strobl (2001) for a meta-study with a similar conclusion. Using administrative panel data on Hungarian firms between 1992 and 2003, Görg et al. (2009) do "not find any evidence for positive horizontal productivity spillovers from foreign affiliates to domestic firms" [...] "on average" (ibid., p. 18). In their anal-ysis, they probe deeper into determinants of potentials for technology spillovers and distinguish between different production technologies or investment motives: they find that FDI projects that exploit labour costs advantages or produce for the local market are unlikely to generate horizontal spillovers, whilst others with a more capital-oriented production technology are more likely to do so. The domestic firms that are then the main beneficiaries are rather large. They also find that positive technology transfer between foreign affiliates and domestic firms of the same industry had been stronger in the early periods of transition, whereas in later phases the negative competition effect dominated.

Torlak (2004) examines the five transition countries of the Czech Republic, Poland, Hungary, Romania, and Bulgaria using panel data on more than 8,000 plants to conclude that productivity spillovers are either negative or insignificant. Konings (2001) establishes in an analysis of Poland, Bulgaria, and Romania that there is no evidence of positive horizontal spillovers, but all the more negative spillovers are found in the cases of Bulgaria and Romania.

Djankov and Hoekman (2000) find negative effects of FDI presence on domestic firms of the same industry for the Czech Republic and the years 1992–1996, which they explain as a result of the competition effect. Interestingly, the lower the foreign ownership share the more this negative effect loses significance. Likewise Kinoshita (2000): by analysing firm-level panel data in Czech manufacturing between 1995 and 1998, she establishes that there are no technological spillovers from having a foreign joint venture partner (and yet, she finds positive spillovers in one particular high-tech innovative sector).

Jensen (2002) analyses the Polish food industry sector to find no evidence for horizontal spillovers. Zukowska-Gagelmann (2001) concludes, in her comprehensive study on productivity spillovers from FDI projects in the Polish manufacturing industry, that there are negative productivity spillovers, especially on private firms in highly competitive sectors. Bosco (2001) finds for Hungary that spillovers are elusive, no general conclusions are possible, and that spillovers – if at all – can only be found in selective high-tech industries.

By contrast, Sinani and Meyer (2004), testing FDI in the case of Estonia between 1994 and 1999 by use of a production function, find large and positive spillovers which depend in particular on absorptive capacities at the receiving end and on the technology gap between domestic and foreign firms. Moreover, they are able to establish a positive competition effect.

Likewise, Smarzynska Javorcik and Spatareanu (2003) investigate in a panel of Romanian firms (from the Amadeus database) during 1998 to 2000 whether the existence of spillovers is affected by the foreign ownership share. Their analysis finds positive intra-sectoral spillovers, but only for fully foreign owned subsidiaries, whereas joint ventures (mixed foreign/domestic firms) produce no positive horizontal spillovers.[4] This is obviously in direct opposition to the results generated by Djankov and Hoekman (2000). Their analysis is hence widely cited.

Békés et al. (2009), using firm-level Hungarian data for 1992–2003 and the growth accounting TFP-method, find significant positive spillovers that mainly depend on firm size (positive) and productivity gaps (negative): the most productive domestic firms are most likely to benefit from spillovers whilst the least productive firms are affected negatively (again: absorptive capacity). Sgard (2001) also finds significant positive spillovers in Hungary in the very early transition phase during 1992–1999, but only where FDI is associated with export orientation and geographically divergent (with larger spillovers in the northeastern part of the country, between Budapest and the EU border, and lower in the less developed regions, a result probably related to spatial proximity in clusters and the size of the productivity gap and absorptive capacities). The analysis uses a large database of Hungarian enterprises, representing close to 90% of the manufacturing and construction sector. Schoors and van der Tol (2002) likewise find positive evidence of spillover effects both within sectors and between sectors by researching Hungarian company-level data from the Amadeus database.

Kolasa (2007) uses firm-level panel data covering the Polish corporate sector and finds horizontal spillovers. His analysis also confirms the important role of absorptive capacities. Dries and Swinnen (2004) establish for the Polish dairy sector positive horizontal spillovers which lead to improved access to finance, increased investments and quality improvements by small local suppliers.

There are many more examples in the literature, and it would be not only unrewarding but actually impossible to present a complete representation of the literature here. In sum, the review of literature may arrive at the same conclusion as a meta-analysis of horizontal spillovers in developing countries (Wooster and Diebel, 2010, p. 652):

> the empirical evidence, taken as a whole, provide[s] weak support of the existence of gains to domestic firms in developing countries from the presence of FDI in the same sector. Individual study results are highly sensitive to model specification and studies that use panel and firm-level data tend to find insignificant or even significant negative horizontal FDI spillovers. [...] Based on the empirical evidence, therefore, it is quite possible that intrasectoral spillovers from FDI in developing countries are largely nonexistent.

It is clear that this ambiguity in empirical literature is a result of a lack of a comprehensive theory that would describe spillovers as a necessary phenomenon under a given set of conditions (ibid., p. 652). As mentioned earlier, this is not the case, and the root of the problem lies in heterogeneity on behalf of foreign affiliates, foreign investors, domestic firms, and the host country economic and political environment. The more heterogeneity is accounted for, the more empirical analyses tend to be able to find significant results.

6.1.3 Empirical analysis of vertical spillovers in CEECs

With a view on the issue of protection of ownership advantages with its dampening effect on mainly horizontal spillovers to host economy competitors, vertical spillovers should be stronger and hence easier to identify in empirical analysis: vertical, and in particular upstream spillovers in less developed and emerging host economies will typically be intentional in an attempt to upgrade supplies. Downstream spillovers are equally intentional where marketing outlets are upgraded to suit the quality expected of the brand name (for example automobile dealers, franchising, see for example Altenburg, 2000), where industrial customers buy machinery, or where FDI is in the infrastructure industry and provides new or better services than before in the local economy (for example Meyer, 2003, p. 24).

Vertical spillovers are typically associated with linkages, themselves measured in shares of output of total industry in the host economy sold to or procured by foreign owned firms. This is done by using input–output tables. Vertical linkages are then used in the usual production function approach to estimate their role in productivity spillovers.

Whilst Békés et al. (2009) were able to identify significant horizontal spillovers in Hungary for 1992–2003, the panel regression analyses do not identify significant vertical spillovers, neither for backward nor for forward industries (only where foreign affiliates are exporters are backward spillovers positive and significant). Schoors and van der Tol (2002) find positive spillovers for upstream industries but negative ones for downstream industries also on the case of Hungary. In either case, openness in the form of exposure to foreign trade makes spillovers in both directions particularly pronounced.

Damijan et al. (2003b) find support for vertical spillovers in the Czech Republic, Poland, and Slovenia for 1995–1999, whereas in Bulgaria only foreign affiliates benefit from backward spillovers, and in Lithuania and Latvia such spillovers even turned out to be negative. Still, where positive backward spillovers were detected, they were on average ten times larger than the corresponding horizontal effects. In Damijan et al. (2003a), no vertical spillovers, whether backward or forward, are found.

Smarzynska Javorcik and Spatareanu (2003) find significant positive spillovers for upstream industries in general, but they also find significant but negative spillovers where those are fully foreign owned. Their explanation for this holds that greenfield investment projects are typically less likely to source locally due to the costs associated with finding domestic suppliers – little commercial interaction obviously also effects little potential for technological spillovers. In her single-authored study of Lithuania, Smarzynska Javorcik (2004a) finds positive productivity spillovers for local suppliers in upstream sectors (which become insignificant where full ownership foreign affiliates are considered), whilst forward linkages remain insignificant.

Kolasa (2007) finds positive spillovers to local firms in upstream industries for Poland between 1996 and 2003, whereby the size of backward spillovers turns out to be much larger than the size of horizontal spillovers. In the analysis, forward inter-industry spillovers remain insignificant throughout all model specifications.

The purpose of this selective review is to show that the empirical literature on spillovers remains rather inconclusive, even if dealing with only a small region and a short period of analysis: some analyses are able to identify significant positive spillovers, and others negative spillovers, and again others find no significant effects. Where spillovers are found, however, vertical backward spillovers to supplier firms in upstream industries seems to be the largest (for example Smarzynska Javorcik, 2004a; Jindra, 2006, pp. 36–50, 61–65) and dominate both downstream or horizontal spillovers. The main reasons to be found in the empirical literature for the elusive character of spillovers include most importantly (i) the assumption that spillovers may not exist to a larger extent because of a lack in absorptive capacities, and (ii) the large extent of heterogeneity, so that spillovers may only be measurable for small sub-sets of firms or even for individual firms with their own individual characteristics (geographic distance, time/dynamic character, heterogeneity of foreign affiliates, as for example mandates, autonomy, heterogeneity of foreign investors, as for example strategic motives, heterogeneity in the host economy, as for example regulatory regimes including competition policy and IPR, industrial structure) (for the latter, see also Lipsey and Sjöholm, 2005), and (iii) the possibility that aggregating firms into averages may distort the picture in that positive spillovers for some firms are cancelled out by negative spillovers for other firms. Still, the expectation of finding spillovers is rooted in their prediction in theories and concepts of international business (and not least in the wishes and expectations of policymakers in host countries). In any case, it is heterogeneity that in many ways causes problems for the empirical analysis of spillovers.

The analysis presented hence explicitly considers as many sources of heterogeneity as practicably possible. The empirical analysis is again focused on the determinants of spillover potentials rather than an attempt to measure the significance and size of such flows of knowledge and technology. The determinants that emerge from the literature review (foreign investment strategies, investment motives, interaction (and local proximity) between foreign affiliate and domestic firms, degrees of foreign ownership, R&D capabilities of domestic firms, and firm size) are largely the same as the determinants of the opposite direction of flows analysed in Chapter 5 on the role of the host innovation system for the foreign affiliate's own technological activity. Hence the analysis here can use largely the same data as did that chapter.

6.2   The data used for the analysis and testable hypotheses

The insights summarised in the review result from empirical analysis which is guided by theories and concepts prevalent in the international business literature and where it is concerned with the diffusion of knowledge and technology via inward FDI (the spillover literature). The basic assumption is that the developmental impact of foreign affiliates in their host economies or regions increases with the potentials for knowledge and technology diffusion.

Lacking direct information in the IWH FDI Micro Database about the extent of spillovers, the following analysis is concerned with identifying the determinants of spillovers. The potentials for spillovers are proxied in the following analysis by the foreign affiliate managers' estimation of their own importance as a source of technological knowledge for R&D or innovation for local suppliers (vertical upstream), local customers (vertical downstream), or local competitors (horizontal). The IWH FDI Micro Database of 2006–2007 contains a number of variables that can be used in an analysis of conditions testing for spillover potentials. It contains information from 144 foreign affiliates in Croatia, 295 in East Germany, 110 in Poland, 220 in Romania, and 40 in Slovenia.

In particular, the database offers information about the level of importance that foreign affiliates' managers attach to their own firms as a source of technological knowledge for R&D or innovation for local customers, for local suppliers, and for local competitors in the same relevant industry[5]. This information is ranked on a Likert scale ranging from 1 for "not important" to 2 for "little importance", 3 for "important", 4 for "very important", and 5 for "extremely important".

Because in this analysis, all firms are (potentially) included in the analysis and no restriction is applied to technologically active firms, the analysis also accounts for the intensity of technological activity of the foreign affiliate. This follows the plausible assumption that even non-innovative and non-R&D firms may have some knowledge and technology that is relevant to competitors, suppliers and customers (for example if it was received in embodied technology from the foreign investors' network). Technological activity is proxied by an unweighted average of product and process innovation intensities of the foreign affiliate[6]. Managers of innovative foreign affiliates were asked to evaluate the intensity of their product and their process innovations during the three years between 2002 and 2005 in comparison to their competitors in the relevant market (again, innovations were defined as being new to the firm and not necessarily new to the market). Based on the two Likert rankings of the individual indicators ranging from 1 to 5 with increasing comparative intensities, *inno* is, then, a non-continuous discrete indicator ranging from 1.0 to 5.0 in 0.5 intervals with the intermediate levels of 1.5 for "very low to below average", 2.5 for "below average to average", 3.5 for "average to above average", 4.5 for "above average to very high".

Assuming that spillovers will (at least to some extent) depend on the foreign affiliate receiving knowledge and technology from its investor, a proxy for internal technology transfer is used. The database contains a direct question to this effect. To get plausible answers in the fieldwork and to make it easier to estimate a value, the information requested was further specified to include only technology that is embodied in products that the foreign affiliate already produces without substantial adjustments. Some of the technology and knowledge may be tacit, making it even more difficult to assess internal direct technology transfer. The solution of measuring the importance of technology embodied in products for own technological activity accounts for these problems. Restricting the definition of direct technology transfer to products already produced by the foreign affiliate (without substantial adjustments) further narrows down what may be considered technology transfer to something that already comes close to hard facts for the foreign affiliate manager[7]. The resulting proxy can be assumed to contain a downward bias, underestimating the correct extent of direct technology transfer. For the econometric analysis using a probit estimation technique, however, the actual level is less important than the more critical and questionable assumption that the extent of underestimation is the same across all firms. Of the total number of foreign affiliates (n=809), 704 provided an estimate of the importance of this direct technology transfer for their own technological activity. The proxy *tt.mnc* is ranked on a Likert scale ranging from 1 for "not important" to 2 for "little importance", 3 for "important", 4 for "very important", and 5 for "extremely important".

Also new to this analysis is a proxy of absorptive capacity of other firms in the host innovation system, again measured in the perception of foreign affiliate managers'. The database does not contain any information about domestic firms, and localised information on R&D amongst the actors in the host innovation system that is comparable across countries is still very scarce for CEECs in official statistics. Therefore, the analysis uses the indicators in the database that describe the estimations of foreign affiliates on the importance of the host innovation system for their own technological activity. This is close enough to describe the quality of the host innovation system, and has the advantage that this quality is measured in the perception of foreign affiliate managers, the relevant persons when enquiring into the quality of the host innovation system for FDI affiliates. The data used is the same as that used in Chapter 5 as dependent variable, that is, the importance of local scientific institutions, local suppliers, and local customers for foreign affiliates' own technological activity (see Figure 5_11 on Averages of levels of importance of host innovation system actors for foreign affiliates' technological activity). Because data for local suppliers and customers is not available for the Croatian and Slovenian sub-panels, the proxy is split into two alternatives, the absorptive capacity of local scientific institutions (*tech_abil_si*) and the average levels of importance attached to all three actors in the host innovation system (*tech_abil_host*). Both are tested in the empirical analysis.

The proxies for the two strategies of 'home-base augmenting' and 'home-base exploiting', the indicators for the five strategic investment motives, the extent of trade with local suppliers and customers, the variable for autonomy, foreign affiliate age and size, the dummy for greenfield investments, and the dummies for host countries and industries are all identical to the ones used above.

What cannot be quantified in the analysis of the selection of five Central East European economies are variables controlling for regional characteristics and for spatial proximity, as the database does not contain information about the exact location of the foreign affiliate (the confidentiality issue explained above).

The available variables and proxies allow testing a set of ten hypotheses on determinants of knowledge and technology diffusion from foreign affiliate to the host economy:

H(1)   A foreign affiliate that has directly received knowledge and technology from its foreign investor is more likely to diffuse knowledge and technology to the host economy.

H(2)   A foreign affiliate following an augmenting strategy[8] is less likely to diffuse knowledge and technology to the host economy.

H(3)   A foreign affiliate following an exploiting strategy is more likely to diffuse knowledge and technology to the host economy.

H(4)   The more intense the technological activity by the foreign affiliate, the more likely it is to diffuse knowledge and technology to the host economy.

H(5)   The higher the autonomy of a foreign affiliate, the more likely it is to diffuse knowledge and technology to the host economy.

H(6)   The higher the absorptive capacities of an actor in the host innovation system, the more likely the actor is to benefit from knowledge and technology diffusion from the foreign affiliate.

H(7)   The higher the share of local supplies of a foreign subsidiary, the more likely is the diffusion of knowledge and technology to local suppliers.

H(8)   The higher the share of local sales of a foreign affiliate, the more likely is the diffusion of knowledge and technology to local customers.

H(9)   The longer the foreign affiliate is established in the host economy, the more likely it is to diffuse knowledge and technology to the host economy.

H(10) The larger the foreign affiliate in terms of employment, the more likely it is to diffuse knowledge and technology to the host economy.

The analysis below includes all CEECs in the database by use of information from the IWH FDI Micro Database. Whilst all firms – not only innovative ones – are included in the analysis, it has to be noted that many firms drop out in this analysis due to missing values. Not all firms have answered all questions, and the effect of an unbalanced sample is particularly large in this part of the analysis.[9]

6.3 The importance of foreign affiliates for technological activity in the host innovation system

6.3.1 Descriptive statistics

The question of whether foreign affiliates are important sources of knowledge and technology for other local firms in the host innovation system (that is, whether local firms in the host innovation system benefit technologically from foreign affiliates in their region), is tested here by focusing on three main actors: local competitors, local suppliers, and local customers.

Figure 6.2 shows that there is large variation in the opinions of managers of foreign affiliates between countries, but not so much between the three potential beneficiaries The main inter-country differences occur for a higher level attached to customers than to competitors or suppliers in the case of East Germany and Slovenia (even though in the latter, no comparison can be made between suppliers and customers), and in Poland, it is suppliers who receive a much lower level than either competitors or customers. The higher levels attached to customers suggests that where FDI produces prefabricated products, their supplies to downstream customers either have important embodied knowledge and technology or require knowledge and technology for their use that

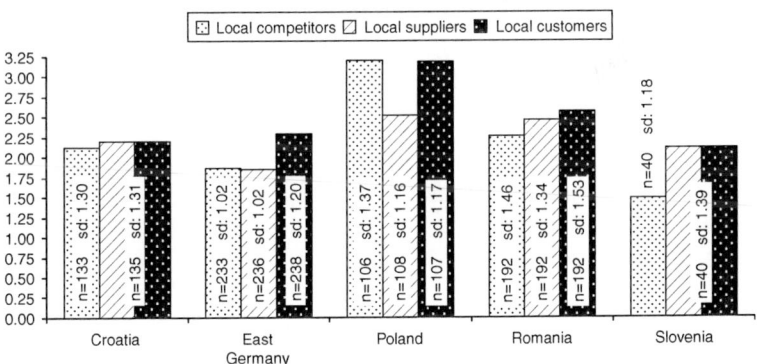

Figure 6.2 Importance of foreign affiliates for technological activity in the host innovation system

Note: n is the number of firms that have provided information; in the Slovenian and Croatian subsamples, the values for customers and suppliers are identical because they had been interrogated as one group. sd denotes standard deviation.

Source: IWH FDI Micro Database 2006–2007.

is supplied by the upstream foreign affiliates. High levels attached to suppliers would have meant that upstream produces of prefabricated products used by the foreign affiliates are involved in a process of technological upgrading by foreign affiliates. Downstream technology spillovers, however, appear to dominate the scene amongst all country groups in the database, while potentials for upstream technological upgrading are portrayed as less important (again, this distinction cannot be made for Croatia and Slovenia due to lack of data). In the analysis of importance of actors in the host innovation system for the foreign affiliate's own technological activity, it was also customers, that by and large dominated the scene, not suppliers (see Figure 5_11 Averages of levels of importance of host innovation system actors for foreign affiliates' technological activity): this does suggest that the world of suppliers in CEE still appears to be of less relevance for FDI projects). This is at variance with the result generated from related research and reviews that spillovers to upstream industries are more easily identified than spillovers to downstream industries (see the review above in this chapter).

High levels of importance attached to competitors are good news for the host innovation system, but may be seen as a potential risk by foreign affiliates housing investor-specific knowledge and technology, due to unwanted spillovers and the risk of supporting future potential competitors. Here, the strength of the IPR regime will determine whether or not potential risks will lead to an exclusion of networks of potential competitor. Only in Poland is the level for local competitors comparatively high (both with respect to a comparison between countries and between the three actors).

In a comparison between all countries, it is Poland's local host innovation systems that is thought to be the most inclined to benefit from local foreign affiliates. This is particularly strong for local competitors and local customers. In the other countries, the average levels of importance are more similar between the three actors in the host economy, and at clearly lower levels. Whether this is a robust result or whether this reflects country-specific perceptions about oneself as a source of knowledge for the technological activity of others remains unknown. Whilst the country-comparative results of Chapter 5 did at times come as a surprise (as they did not reflect the usual pattern of relative states of transition and levels of economic development), the result for the opposite direction of technology spillovers is not at all surprising and reflects the relative levels of technological expertise between domestic firms and foreign affiliates. Those may be particularly high in countries with medium levels of economic development. Poland is sufficiently developed to attract technological FDI whilst the Polish regional innovation system is 'sufficiently less technologically advanced' to let its actors significantly benefit in terms of technology spillovers. It is hence no surprise that the levels of East Germany, Romania, and Croatia are lower: in the case of Germany, the hosting innovation systems are already technologically sufficiently advanced to reduce the potentials for technology spillovers (low technology gaps); in the latter two countries, the host country still appears less attractive to technologically advanced FDI projects, which also reduces the potentials for spillovers.

As in the previous chapter, the country-specific results do already convey a tentative picture, but further analysis is necessary to dig deeper (note also that the standard deviations are quite high, with around 40 to 70 per cent of the corresponding variables, which reduces the robustness of the messages they appear to convey). This again has to consider heterogeneity at the micro-level, which is possible by a firm-level multivariate regression analysis using the IWH FDI Micro Database. This time, the potentials for spillovers assume the role of dependent variables, and the econometric models test the role of various determinants, some of which had already been used in the previous analysis, in Chapter 5). Several econometric tests are conducted to evaluate the hypothesis generated above, and for which the database offers relevant proxies. This helps to find answers to the research question of this chapter about the determinants of a positive role of inward FDI for the technological activity of the host innovation systems in CEECs. All tests are again ordered probit specifications, reflecting the character of the data underlying the analyses.

As independent variables for the econometric tests, the analysis again considers the two strategies of *augm, expl*, the local embeddedness in terms of trade *loc.sales, loc.suppl*, the autonomy issue with the variable *autono3, age*, and *size*, the five strategic motives, the dummy for greenfield investments, and the dummies controlling for industries and host countries.

In contrast to the previous tests, the regression analyses in this chapter include all data, and not only that for innovative firms. New to the analysis here is, hence, an indicator of technological activity of foreign affiliates: intensity of innovative activity *inno* for product and process innovation in comparison to their competitors in the relevant market. Figure 6.3 presents the average innovation intensities of foreign affiliates according to their host countries.[10]

The messages conveyed by the amalgamated variable *inno* leaves the ranking implied by the individual variables for product and process

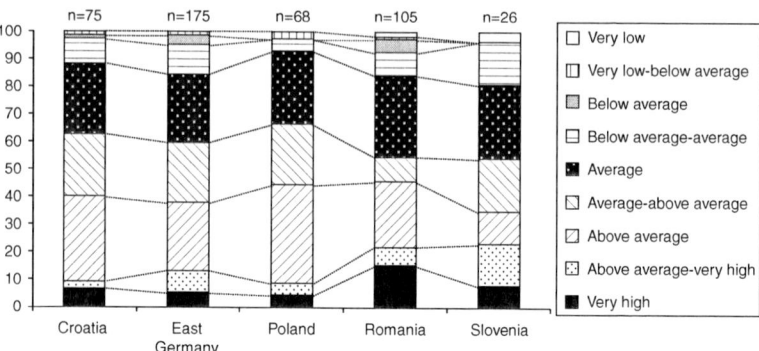

Figure 6.3   Averages of product and process innovation intensities in comparison to competitors (*inno*), by host country

Note: Averages between product and process innovations are computed as simple unweighted averages between the intensities of product and process innovations. n is the number of firms that have produced product and process innovations during the years of 2002–2005 and have provided information on the intensity. sd denotes standard deviation.

Source: IWH FDI Micro Database 2006–2007.

innovations by and large unchanged: when using the share of foreign affiliates considering their innovative activity as "average to above average" as the cut-off point *vis-à-vis* everything at average levels or below, then Poland remains the country sample with the highest (perceived) intensities of innovation, followed by Croatia and East Germany; Romania and Slovenia remain at the bottom of the list. If the cut-off point is changed to "above average", then Romania moves up to the first rank, together with Poland.

Additionally, the empirical analysis includes a proxy for the extent to which the foreign affiliate has received technology from MNC group, here specified as technology that is embodied in products that the foreign affiliate already produces without substantial adjustments (*tt.mnc*). Figure 6.4 depicts those estimations for each of the host countries in the database: Polish foreign affiliates on average consider technology received from the parent network to be between "important" and "very important" with an average country-level on the original scale of 3.64. East German affiliates, in contrast, have on average the lowest estimations on the importance of foreign investors' embodied technology (an average level of 2.48). The averages for the other host countries are somewhere in between, with 3.13 for Croatia, 2.87 for Romania, and 2.82 for Slovenia.

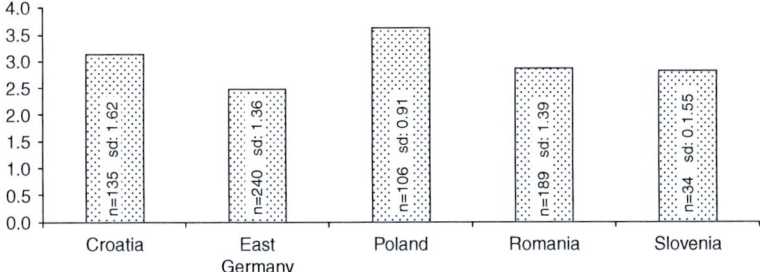

Figure 6.4 Importance of embodied technology received from the foreign investors' networks, by host country

Note: n is the number of firms that have provided information. sd denotes standard deviation.

Source: IWH FDI Micro Database 2006–2007.

Taking this variable as an indication of technology transferred from the parent network to the foreign affiliate and negating possible country-specific differences in perceptions of managers uncontrolled for, then the conclusion would be that Polish foreign affiliates benefit the most from direct, internal technology transfer, followed by Croatian, Romanian, and Slovenian affiliates. Unfortunately, however, standard deviations are sizeable, with between 25 per cent of the average in the case of Poland and as much as 52–55 per cent in the other cases.

The two proxies of absorptive capacities of the local host innovation system (the first on local scientific institutions: *tech_abil_si*, and the second on customers and suppliers: *tech_abil_host*) are presented in Figure 6.5 (reproducing the levels of importance attached to scientific institutions: *tech_abil_si* from Chapter 5 plus the new average indicator of all three actors in the host innovation system: *tech_abil_host*). The first has a higher number of observations and yet only represents scientific institutions; the other is more comprehensive in covering all three actors, and yet has a much lower number of observations, excluding Croatia and Slovenia.

Interpreting this indicator as quality of the host innovation system from a technological perspective and hence as a measure of absorptive capabilities, then Poland and Croatia would offer the most interesting host regions, closely followed by East Germany, whereas Romania and Slovenia rank markedly lower. For *tech_abil_host*, standard deviations are

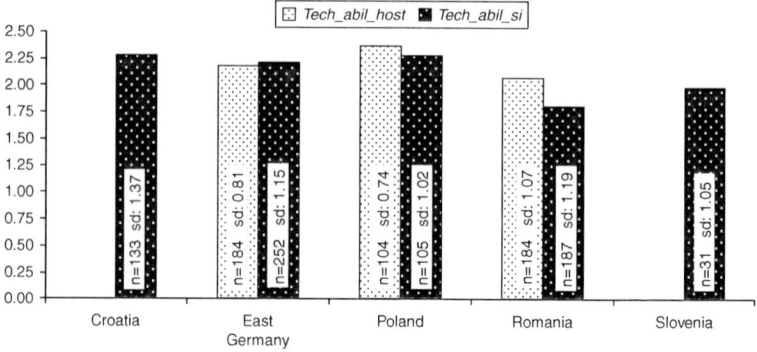

Figure 6.5   Technological quality of the host innovation system from the perspective of foreign affiliates, by host country

Note: n is the number of firms that have provided information. sd denotes standard deviation.

Source: IWH FDI Micro Database 2006–2007.

comparatively low for Poland and East Germany with between 32 and 37 per cent of averages, but 52 per cent in the case of Romania. Standard deviations for *tech_abil_si* are higher, with levels between 45 and 52–53 per cent for Poland and East Germany or Slovenia respectively, and over 60 per cent for Croatia and Romania. The latter averages on scientific institutions hence have to be interpreted with particular care (and the empirical analysis below in fact also turns out to be more robust when using *tech_abil_host*).

The descriptive data again shows that firm heterogeneity exists next to host country differences. This will be accounted for by a firm-level multivariate regression analysis below.

6.3.2   The method of empirical analysis

The hypotheses developed above are tested by use of 11 determinants. With a view on firm heterogeneity, the analysis again controls for investment motives, greenfield vs acquisitions, the industry of the foreign affiliate, and the host country. The dependent variable is the proxy for spillovers, the importance of the foreign affiliate as a source of technological knowledge for R&D or innovation for local firms (divided into customers, suppliers and competitors). The method of analysis is again an ordered probit regression, which is determined by the shape of the dependent variable.

The model is specified with a parameter vector including:

DIFF = tt.mnc; augm; expl; inno; sal.inno; exp.tech; autono3; tech_abil; loc.sales; loc.suppl; age; size; d_market; d_client; d_effic; d_nat.res; d_ know; d_greenf; d_indus; d_host

with *DIFF* being the proxy for the diffusion of knowledge and technology from the foreign affiliate to local firms (importance of foreign affiliates as a source of technology and knowledge for the host innovation system); *tt.mnc* being internal technology transfer from parent to affiliate (importance of foreign investor's network for foreign affiliate's own technological activity); *augm* and *expl* the two Marchian strategies of 'exploiting' and 'augmenting'; *inno* the intensity of product and process innovative activity; *sal.inno* being the share of sales of new products; *exp.tech* the share of R&D and innovation expenditure in total sales; *autono3* the degree of autonomy in the three technology-related business functions; *tech_abil* being the technological ability of the host innovation system (in the two alternative specifications *tech_abil_si* and *tech_abil_host*); *loc.sales* the share of sales to the local market; *loc.suppl* the share of supplies procured on the host markets; *age* in years since the foreign investment was made and as of 2005; *size* in employment numbers; *d_market* as the dummy variable controlling for the investment motive of access to markets that assumes a value of 1 where the foreign affiliate manager values this motive as "important", "very important" or "extremely important"; *d_client* controlling for following foreign key clients; *d_effic* for increasing efficiency across the foreign investor network; *d_nat.res* for accessing location-bound natural resources; and *d_know* for accessing location-bound knowledge, skills, technology.[11] *d_greenf* controls for FDI projects involving new firms; *d_indus* controls for the 14 NACE two-digit single main industrial occupation of the foreign affiliate; *d_host* for the five of the countries of location of the foreign affiliate in Central East Europe (Croatia, East Germany, Poland, Romania, Slovenia).

Before the regression models are tested, multicollinearity between determinants is checked by a pairwise correlation of all independent variables. Table 6.1 shows that none of the significant correlations are in fact at levels that would raise problems for the regression analysis. The highest and significant correlation coefficients are between direct technology transfer and the exploitation strategy with a significant coefficient of .46, which is not surprising at all. The other correlations remain qualitatively the same as in the previous chapter (differences emerge

Table 6.1 Pairwise rank correlation between all determinants

	tt.mnc	augm	expl	inno	sal.inno	exp.tech	autono3	tech_abil	loc.sales	loc.suppl	age	size
augm	.24*											
exploit	.46*	.40*										
inno	.22*	.18*	.11									
sal.inno	.06	.19*	.04	.15*								
exp.tech	-.08	.08	.05	.11	.15*							
autono3	-.23*	-.15*	-.44*	.11	.05	.11						
tech_abil	.15*	.20*	.22*	.18*	-.06	.05	.09					
loc.sales	.00	-.04	-.11	-.06	-.01	.21*	.11	.20*				
loc.suppl	.08	.18*	.08	.08	.05	.06	.20*	.24*	.36*			
age	.05	-.01	.02	-.09	-.08	-.01	-.08	.01	.10	.04		
size	.14*	.01	.12	.04	.09	.18*	.15*	.09	.23*	.18*	.02	
d_greenf	.12	-.05	-.01	-.01	-.14*	.05	-.17*	-.23*	-.11	-.12*	.18*	-.12*

Note: tech_abil was tested in both specifications (_host and _si). The results presented here are for the _host specification, because the subsequent analysis shows that this specification produces more robust results. All correlations were conducted as Spearman rank correlations using casewise deletion, where observations are ignored if any of the variables in the list of variables are missing: obs=198. Significant correlations (at a level of .1) are marked with an asterisk. Correlations between dummies for investment motives are not reported due to restriction in space: the coefficients all remain lower than .3.

Source: IWH FDI Micro Database 2006-2007.

only due to the lower number of observations here).[12] The correlation between the intensities of product and process innovation is also high and significant, and in terms of hypotheses a differentiation between the two is not necessary, hence the use of the average between the two.

Some of the methodological caveats of such an analysis as discussed in the previous chapter likewise apply to these analyses here: the data used is subjective, and there is some degree of eclecticism in the theories or concepts underlying the analyses. Therefore, again several model specifications are tested in an attempt to increase robustness of interpretation of results: model 1 uses all the data available, model 2 omits the five investment-motive dummies, model 3 the industry dummies, model 4 leaves out the country dummies, and model 5 does not consider any of those dummies. All models are tested twice: once for the specification *tech_abil_si* and once for *tech_abil_host*. The results from those two latter specifications are roughly the same, albeit with lower pseudo R^2 for the variation using *tech_abil_si*. In the following, hence, the results pertaining to the specification using *tech_abil_host* are presented. Where interesting differences emerge, those are shown in endnotes.

6.3.3 Estimation results

All models tested are significant overall, Wald tests confirm that all independent variables are significantly distinct from 0 (see Table 6.2). All models have low levels of pseudo R^2 which, again, is typical for ordered probit models with fieldwork data. It is noteworthy, however, that the analyses in this chapter produce generally higher levels of R^2 than that in Chapter 5. In particular in the analysis using model 1, the level here is more than double in size. One explanation is that more determinants are included in the analysis of this chapter (18 instead of 15).

In contrast to Chapter 5, however, the econometric tests show a lower number of significant results and also less consistency between the three actors of the national innovation system. This already suggests that the interpretation of hypotheses will have to be specific to each innovation system actor.

As was the case in Chapter 5, it is again the model that includes all determinants and dummies that performs best in terms of pseudo R^2. This suggests that country dummies and industry dummies do in fact play a role, and controlling for them is important in this analysis as well. In the subsequent analyses testing the set of ten hypotheses, the specification of model 1 is used.

Table 6.3 presents the results for three actors: local suppliers, local customers, and reproduces the results for local competitors for easier

Table 6.2 Determinants of foreign affiliates assuming an important role in the technological activity of firms in the host economy

	Model 1	Model 2	Model 3	Model 4	Model 5
tt.mnc	-.01 (.916)	.00 (.960)	.05 (.514)	.12 (.128)	.19* (.007)
augment	.21* (.032)	.22* (.022)	.15 (.136)	.12 (.204)	.07 (.427)
exploit	.10 (.315)	.11 (.287)	.07 (.513)	.12 (.246)	.09 (.372)
inno	-.05 (.615)	-.02 (.836)	-.04 (.684)	.01 (.890)	.03 (.767)
sal.inno	-.00 (.594)	.00 (.802)	.00 (.444)	-.00 (.975)	-.00 (.929)
exp.tech	-.00 (.740)	-.01 (.522)	-.00 (.567)	-.00 (.342)	-.01 (.165)
autono3	.00 (.951)	.00 (.791)	.00 (.896)	-.00 (.721)	-.00 (.946)
tech_abil[13]	.50* (.000)	.50* (.000)	.46* (.000)	.47* (.000)	.42* (.000)
age	.00 (.960)	.00 (.909)	.00 (.878)	.00 (.853)	.01 (.603)
size	-.00 (.131)	-.00 (.115)	-.00 (.210)	-.00 (.354)	-.00 (.403)
d_market	.26 (.181)	–	.21 (.212)	.30 (.106)	–
d_clients	-.24 (.315)	–	-.20 (.360)	-.17 (.442)	–
d_effic	-.15 (.400)	–	-.06 (.721)	-.10 (.572)	–
d_nat.res	.44* (.064)	–	.30 (.179)	.47* (.033)	–
d_know	-.02 (.938)	–	-.09 (.635)	-.20 (.326)	–
d_greenf	-.11 (.575)	-.13 (.504)	-.18 (.341)	-.07 (.727)	-.19 (.293)

d_indus	yes	yes	–	yes
d_host	yes	yes	yes	–
n	189	189	189	189
Wald χ^2	97.62	94.32	69.75	78.54	36.84
Prob > χ^2	0.000	0.000	0.000	0.000	0.000
Pseudo R^2	0.172	0.159	0.136	0.135	0.091

Note: Coefficients with * and in bold are significant at the .1 level. Figures in brackets denote heteroskedasticity-robust $P > |z|$.

Source: IWH FDI Micro Database 2006–2007.

Table 6.3   Determinants of foreign affiliates assuming an important role in the technological activity of firms in the host economy

	Local competitors	Local suppliers	Local customers
tt.mnc	−.01 (.916)	**.37* (.001)**	.07 (.451)
augment	**.21* (.032)**	**.19* (.084)**	.01 (.909)
exploit	.10 (.315)	.07 (.610)	.05 (.697)
inno	−.05 (.615)	.06 (.650)	.08 (.505)
sal.inno	.00 (.594)	**.01* (.043)**	.00 (.363)
exp.tech	−.00 (.740)	.01 (.511)	−.01 (.177)
autono3	.00 (.951)	**.01* (.027)**	.00 (.838)
tech_abil_host[14]	**.50* (.000)**	**.37* (.018)**	**.55* (.000)**
loc.sales	–	–	**.01* (.000)**
loc.suppl	–	**.01* (.003)**	–
age	.00 (.960)	**.04* (.087)**	.01 (.779)
size	−.00 (.131)	−.00 (.194)	**−.00* (.037)**
d_market	.26 (.181)	−.20 (.379)	.04 (.860)
d_clients	−.24 (.315)	−.07 (.800)	−.05 (.839)
d_effic	−.15 (.400)	−.04 (.853)	−.04 (.810)
d_nat.res	**.44* (.064)**	.10 (.728)	.14 (.521)
d_know	−.02 (.938)	.05 (.826)	.14 (.448)
d_greenf	−.11 (.575)	**−.54* (.009)**	-.13 (.463)
d_indus	yes	yes	yes
d_host	yes	yes	yes
n	189	158	200
Wald ^{2}	97.62	98.27	137.48
Prob > ^{2}	0.000	0.000	0.000
Pseudo R^2	0.172	0.229	0.195

Note: The n of tests for customers and suppliers is much lower than for scientific institutions, because data on the importance of customers and suppliers is not available in the Slovenian and Croatian subsamples. Coefficients with * and in bold are significant at the .1 level. Figures in brackets denote heteroskedasticity-robust $P > |z|$.

Source: IWH FDI Micro Database 2006–2007.

comparability. In the models with suppliers and customers as dependent variables, the corresponding determinant of sales or supplies on the host market respectively is included. The table shows mostly insignificant results for determinants: of the ten topical determinants (internal technology transfer, augment, exploit, innovations, sales due to innovations, expenditure in technology, autonomy, host innovation system's technological ability, local sales and local supplies), it is only the proxy for technological ability of the host innovation system that produces consistent and significant results with a plausible positive sign. In the light of theories, these results lend support to the absorptive-capacity issue: only if the actors in the host innovation system are technologically active (and hence important for the foreign affiliate's own technological activity), will those actors have sufficient absorptive capacities to benefit from technology diffused from the foreign affiliate to the host economy. This significantly supports hypothesis H(6).

Technology diffusion appears to depend on direct technology transfer from parent investor to foreign affiliate only in the case of local suppliers; embodied parent investor technology only appears to be relevant for upstream diffusion of technology, not for downstream or for a horizontal direction within the same industry. For local competitors and local customers, technology spillovers from a foreign affiliate do not necessarily originate from the foreign investor (it may constitute disembodied technology owned by the foreign investor, yet the data is unable to tell), but may rather stem from the foreign affiliate's own technological activity. Hence the results only partly support hypothesis H(1), namely only for local suppliers.

With respect to hypotheses H(2) and H(3) on the two strategies of augmentation and exploitation, the analysis does confirm a positive effect of the augmentation strategy on spillovers for local competitors and local suppliers. However, not even in a scenario of home-base augmentation do local downstream consumers appear to significantly benefit from technology diffusion. In the case of the exploitation strategy, none of the results turn out to be statistically significant. Hypothesis H(4), on the role of own technological activity of foreign affiliates for spillovers to the host innovation system, is tested by use of three indicators. Yet only in the case of the indicator of 'share of sales due to innovations' do local suppliers significantly benefit from technology diffusion; in all other cases, the analyses remain insignificant. In the light of theory, this produces a small amount of support to the concept of technological accumulation by Cantwell, whilst not finding anything to refute this approach.

The little support that the analysis finds for the hypotheses H(1) and H(4) is in fact disappointing and yet still well in line with other empirical findings: technology diffusion from foreign affiliates to local firms is difficult to identify, and often even turns out to be negative. If such diffusion can be identified, then it is mainly in an upstream backwards direction, and this is also reflected in this analysis. It is still somewhat disappointing because, regardless of whether knowledge and technology is of parent investor origin or developed by the foreign affiliate itself, the local economy does not seem to significantly benefit more from FDI projects with more technological assets or activity.

Whilst in the analysis in Chapter 5, the determinant of autonomy was significant throughout, it is here only so with respect to local suppliers: H(5) only appears to be supported for upstream technology diffusion, and it does not seems to be in the foreign affiliate's own interest to let customers or even competitors benefit from their technological ownership advantages. Suppliers, however, have to deliver the requested specification and for that, some upstream diffusion of technology and knowledge becomes indispensable.

The indication of the backward-upstream channel of technology spillovers found above is also reflected by the empirically significant and positive results for local procurement, lending support to hypothesis H(7). But support is also found for hypothesis H(8) on local sales, a direction of spillovers that did not receive much empirical support in the tests of the other determinants. The analysis of age for hypothesis H(9) and of size for H(10) produce mixed results, but if significant, the determinant of size turns out to be negative, which also was the case in Chapter 5 even if here at variance with what had been suggested in the literature review. This suggests that technology and knowledge interaction and diffusion in CEE are more likely between foreign affiliates and small firms, and less so with larger units. Often, the former are newly established enterprises, whereas the larger ones are sometimes the results of the privatisation processes (that incidentally often involved foreign capital, suggesting that possibly competition between foreign investors may drive this result). The test of upstream diffusion does find some support for the hypothesis that longer-established foreign affiliates may carry more reputation: diffusion tends to increase with foreign-affiliate age.

In view of the other control variables, the only important result is the negative significance for the diffusion from greenfield investments to local suppliers. This may go some way to support the 'liability of newness' case.

6.4 Summary of main results

At a general level pertaining to any kind of host economy, the literature review finds that external technology transfer effects (spillovers in the wide definition of the term) are found to be much more elusive than internal technology transfer. Where spillovers are identified as significant in relevant literature on CEECs, they in turn are dominated by vertical, backward linkages, whereas positive vertical downstream spillovers to local customers are more rarely established, and positive horizontal spillovers to local competitors in the same industry tend to be very scarce if identifiable at all. Further, FDI projects seeking to exploit labour cost advantages or producing for the local market are less likely to involve horizontal spillovers than FDI projects with capital-oriented production technologies. Moreover, the largest effects are to be expected amongst large FDI projects that started early in the transition process. Further, absorptive capacities and productivity gaps turn out to be important determinants for vertical spillovers. The role of the foreign ownership shares remains unclear, whilst greenfield projects are typically associated to rather small spillovers. It may also be true that spillovers emanate predominantly from foreign affiliates whose investors are both strategically supportive of generating spillovers and experienced in their management, for example in the form of clusters in their home location or the locations of other foreign affiliates in the TNC's network.

My own empirical analysis conducted in this chapter yields several interesting results: in the descriptive analysis, direct technology transfer appears to have been most intense amongst Polish foreign affiliates, followed by Croatian, Romanian, and Slovenian affiliates. East German foreign-owned firms appear to have benefited the least. Also in the descriptive analysis, the absorptive capacities of Poland, East Germany, and Croatia turn out to be highest and hence signal more attractive locations for the home-base augmentation type of FDI projects, whereas Romania and Slovenia rank markedly lower.

With respect to the research question of this chapter (R6), the econometric analysis is able to identify only a small number of significant determinants of spillovers into the host innovation systems in CEECs. In total, the empirical tests in the available literature remain rather inconclusive about the existence, sign, and extent of technology spillovers via inward FDI in CEECs. Heterogeneity prevails. In my own analysis that accounts for some degree of heterogeneity, the test consistently throughout all model specifications shows that absorptive capacities of the actors in the host innovation matter for the potentials of spillovers. Considering that the proxy for absorptive capacities was built here by proxying the importance of the host innovation systems actors for the technological activity of the foreign affiliate, then this further suggests that spillovers depend on the reciprocity of the exchange of knowledge and technology between the foreign affiliate and its host economy. Further, spillovers do not depend on the foreign affiliate itself benefiting from direct technology transfer; this only turned out to be important for spillovers to local suppliers. FDI projects following an augmentation strategy are typically more likely to generate spillovers to the host economy, but only significantly so to local competitors and local suppliers, not to local customers. Exploitation strategies remain not at all important as a determinant of spillovers. In terms of own technological activity on behalf of the foreign affiliate as a determinant of spillovers, the analysis finds little support, and only for local suppliers.

At variance with the general literature, the analysis here suggests (even if only weakly) that it is rather the small foreign affiliates that generate sizeable spillovers, not the large ones.

What is particularly surprising is that the results also suggest that the local host economy does not seem to benefit more in terms of spillovers originating from FDI projects with more technological assets (endowed via internal transfer from foreign investors) or with more affiliate technological activity. Spillover potentials are found for both upstream and downstream industries.

7
The Role of Intellectual Property Rights for Technology in FDI into CEE

Having treated spillovers as a positive condition for the technology-accelerating role of inward manufacturing FDI, and having analysed conditions at the firm level, this chapter now approaches the issue of conditions of technology transfer from a different, complementary perspective: now, the ownership advantage aspect à *la* Dunning of technology and knowledge owned by the foreign investor moves into the limelight. Knowledge and technology constitute firm-specific (ownership) advantages that are increasingly important as a competitive factor between firms in the industrialised world. Knowledge, however, has the particular characteristic that it may be applied by several users at the same time without diminishing its content, possibly even the converse (think, for example, of the accumulative character of an idea). Whilst there is no technical exclusivity for the use of knowledge, the commercial value of knowledge can very well be held to be exclusive: a firm has a competitive advantage over rival firms if it uses a particular piece of knowledge or technology that it exclusively owns and competitors therefore do not have, or cannot, are not allowed to apply.[1]

Firms holding such ownership advantages hence have a motivation to protect their knowledge-related ownership advantages from unwarranted use by others. Such ownership can be protected by keeping secret a particular piece of knowledge that constitutes the technological advantage of the firm (trade secrets). Where the knowledge cannot be protected by secrecy (think of, for example, reverse engineering, easy-to-copy technology), the resulting dissemination process involves negative externalities for the foreign investor and, vice versa, positive externalities for the institutions at the receiving end, and the state

177

governing a market economy is called upon to regulate. The formal institution designed to counteract such market failure is the intellectual property regime; it guarantees legal exclusivity of use to the declared owner of knowledge and technology that is deemed to warrant protection (that is, novel in the sense of not having been commercially applied previously and non-obvious in its providing a solution to a non-trivial problem). This institution is market-supporting in as much as it reduces the risk of a registered innovator of being unable to capture the profits generated by an innovation. This risk may be of a technological nature (ease of copying, imitating), may be related to the characteristics of knowledge (tacit knowledge is more easily retained by an innovator than codified knowledge), or may be of a legal/transaction cost nature: the IPR regime is just that institution. Where firms choose to use this institution to protect their knowledge, the proprietary technology is published in, for example, a patent, together with a description of possible uses of the technology (claims). Other firms, including (potential) rivals, can thereby get access to that knowledge and may be able to use it or fractions of or derivatives of that knowledge to their own profitable advantage. The system of an IPR regime is designed to both (i) promote dissemination of new knowledge and at the same time (ii) offer the owner of a knowledge title the possibility to charge fees for the use of that knowledge (licensing of a patent). The former gives rise to the intended effect of generating a 'market' (see for example Andersen et al., 2010) in which knowledge and technology are tagged with a price and efficiently allocated to the most able user when traded freely. The latter design feature is motivated by the fact that the development of knowledge may have involved costs (for example in the form of R&D), and the IPR regime is the institution that allows the originator of that knowledge to generate profits to make up for this investment. The ability to recoup knowledge-generating costs hence depends on the ability of this institution to protect proprietary knowledge from unwarranted use and to guarantee appropriability – and that ability may well diverge between different designs or intensities of enforcement of IPR regimes across different countries.

When the CEECs joined the EU in 2004, they had all enacted a Western-style IPR regime already, and aligned this to the *acquis communautaire*. The legal framework should hence be nominally uniform between East and West Europe. Having comparable laws on the books, however, does not necessarily translate directly into equal enforcement of the law: the level of protection against abuse of IPRs depends not only on the way the law is interpreted in each individual case (which is

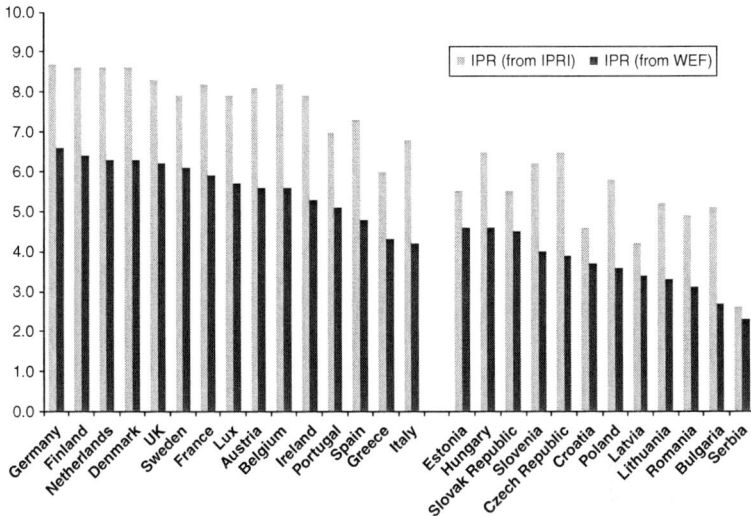

Figure 7.1 IPR indicators in Europe, East and West

Note: The WEF index has a theoretical maximum value of 7.0, whereas the IPRI index is highest at 10.0.

Croatia and Serbia are included as prospective future EU members.

Sources: WEF (2007), Section IX, Table 3.3; IPRI (2009), p. 20.

a political issue) but also on the costs and time involved in lawsuits (the transaction-cost argument).

The level of protection provided by an IPR regime differs markedly between European countries. The most obvious difference in Europe is between the West with higher levels and the East with lower levels. The two most widely used indicators[2], the Intellectual Property Protection index by the WEF (2007) and the "Intellectual Property Rights" section of the International Property Rights Index (IPRI, 2009), clearly rank CEECs lower than West European countries, with only two Western countries overlapping (Greece and Italy rank lower than the three highest CEECs: Estonia, Hungary, and Slovenia) (see Figure 7.1).

Whilst the strictness of IPR protection is undoubtedly closely related to the level of economic development of the country's economy (measured for example in GDP per capita and regardless of whether in Euros or in power-purchasing parities)[3], the direction of causality remains an open issue. Is the effectiveness of an IPR regime a result and a reflection of the level of economic development (for example because enforcement involves administrative costs and significant human capital in agencies and courts)? Or do policymakers in less developed countries intentionally favour a less stringent enactment in an attempt to reduce the costs for domestic enterprises when acquiring and applying protected (foreign) technology?[4]

In any case, though, a policy dilemma emerges: on the one hand, FDI is amongst the most important sources of new technology in particular in emerging markets, and strict enforcement of IPR rules directly raises the costs of international technology transfer. Yet on the other hand, will weak IPR regimes have an effect on either FDI itself, or its technology-carrying propensity, or both, possibly indirectly raising the costs of international technology transfer? The literature assumes that technology effects via inward FDI often fall short of expectations, mainly because foreign investors tend to protect their core knowledge from dissipating to local rivals (see for example Veugelers and Cassiman, 2004).

Because most of the FDI projects into CEECs originate from the West, these projects operate under a distinct gap in the level of IPR protection between home and host country. Other prominent FDI-attracting regions which also have emerging markets rank even lower (for example Russia: 2.4 of max. 7.0 or 4.3 of max. 10.0), but levels in CEECs are well in the range between those in China (3.3 of 7.0, or 4.4 of 10.0) and India (4.5 of 7.0, or 5.1 of 10.0). European East–West IPR gaps are, hence, a relevant matter of concern.

The focus of analysis in this chapter is concerned with the relationship between IPR regimes and the technological role of inward FDI for the host economies of CEE. hence the main research questions to be asked here are:

RQ(7_1)   Will the comparatively weaker IPR regimes in CEECs negatively affect the technological activity of foreign affiliates?

RQ(7_2)   Will the comparatively weaker IPR regimes in CEECs inhibit the technological embeddedness of foreign affiliates into their host innovation systems?

Those two questions stand out as particularly important not least in terms of economic policy aimed at attracting inward FDI: being rather broad-based policy measures that affect both domestic and foreign owned firms alike, IPR regimes prominently belong to the important institutional framework conditions for inward FDI.

Following a brief description of what theory and related empirical analysis may hold on this issue, a set of individual empirical analyses has been devised to contribute to answering these research questions. The empirical analysis is, as in the previous chapter, designed to test hypotheses that had been developed from the review of the current state of the art in research. The analysis, however, follows a different structure: each hypothesis is tested in a separate analysis rather than in the framework of one comprehensive model. This is necessarily due to the particularity of the hypotheses.

7.1 The current state of the art in research

There is a striking gap in the literature where the two issues of TNCs or FDI and IPR merge: so far, the literature is mainly concerned with the kind of foreign investment made in weak IPR regimes (entry modes) and rather little is known about the effect on foreign-affiliate technological activity and the technological embeddedness of FDI in weak IPR systems.

In this chapter, two strands of literature are combined: institutional theory and international business theory. The literature simultaneously concerned with both strands is wide and has produced many important insights into the role of institutions on the process and effects of internationalisation of firms. This is particularly insightful in post- transition economies. Here, profound changes in institutions occurred, and those can be analysed in terms of their effects ('societal quasi-experiment' Meyer and Peng, 2005). Further, formal and informal institutions have often turned out to be insufficiently compatible or even completely incompatible: formal institutions (typically, EU ones) were implanted and could not gradually evolve in harmony with gradually changing norms and values.[5]

7.1.1 Institutions and internationalisation in general

One of the most pertinent issues in the literature on institutions and FDI pertains to the institutional determinants of FDI decisions and entry modes. The relevant forms of internationalisation here include contractual arrangements involving anything from licensing to technology sharing or technology transfer agreements, to equity solutions in joint ventures, to FDI via merger and acquisition or greenfield investment. From the internalisation concept in international business theory based on Coase, and further developed by Williamson's transaction cost theory, we take that costs of transactions on markets may well be significant, and if they are higher than costs of internal coordination then ownership solutions become efficient. In this respect, we find for example that national regulations on foreign investors play a role (for example Contractor, 1990: the more restrictive, the lower the ownership share in joint ventures), or politically induced investment risks (for example Agarwal and Ramaswami, 1992: the higher the risk, the less likely are FDI solutions and the more likely become joint ventures or modes of internationalisation without ownership). With respect to cultural differences between host and home, we find two conflicting suggestions (for example Kogut and Singh, 1988: the higher the cultural distance, the higher the probability to choose joint ventures over acquisition – alas in Shane, 1994: cultural distance is associated with wholly owned FDI modes, see also Brouthers and Brouthers, 2001, with an attempt to reconcile these contradictions by adding investment risks to this issue).

This feeds into the literature on the effects of internationalisation on technology transfer and spillovers. Whilst in general the results remain unclear (see for example Görg and Greenaway, 2003), the more robust results are generated for direct technology transfer through FDI from investor to the foreign investment, and the results for spillovers are much less conclusive. The general picture from the literature is that probably the most significant spillovers can be found to be vertical and upstream within the same industry. Next to the problem of firm heterogeneity, Arora (2009) suggests that one reason for the shortcoming may be rooted in the fact that patent protection and IPR regimes in general differ between the countries and internationalisation projects assessed in the literature.

7.1.2   The IPR regime and internationalisation

The relevant literature concerned with the IPR regime as a particularly important institution for technology transfer and its effect on entry strategies generally supports the plausibility assumption that FDI and the strength of IPR protection will have a positive relationship. Nevertheless, many qualifications of this relationship have been found and are discussed below. At the most general level, empirical analysis has produced results that vary from the research of institutions in general. Those results suggest that the role of institutions for a given subject of research has to distinguish between different forms of institutions, and in particular, their effects may not always be linear (as proposed earlier for different forms of internationalisation by Meyer, 2001, p. 361). On the one hand, we find evidence that the amount of FDI inflows will tend to increase with the strength of IPR protection in the host country (see for example Braga and Fink, 1989, p. 172; Beamish, 1993, and Yan and Gray, 1994, in applications to China;

Lyles and Salk, 1996; and Maskus, 1998). This is largely in line with what has been established for institutions in general: the more developed are the institutions (including IPR regimes), the more likely will international-isation take the form of ownership over contractual forms and foreign trade. On the other hand, however, Park and Lippoldt (2003) find that the positive association between FDI and improvements in IPR protec-tion depends on the starting level of the IPR regime: the reform of an initially lower level of IPR protection leads to a higher rise in FDI. For firms, changes for the better appear to have been more important than the current national state of the art in the strength of the IPR system. With respect to the mode of internationalisation, Buckley and Casson (1996) (see also Shane, 1994) suggest that 'missing patent rights' make licensing less attractive, whilst joint ventures as well as mergers and acquisitions (that is, FDI) become more likely (the possibility of direct control by way of FDI may explain this result). Oxley (1999), by use of transaction costs economics, likewise finds that firms will use more hierarchical alliances (joint ventures rather than contract-based alliances) where IPR regimes offer weak protection. Further, the literature holds that forms of inter-nationalisation that involve less ownership, such as joint ventures and licensing, may take precedence over FDI in environments with stronger IPR systems (for example Nunnenkamp and Spatz, 2003; Braga and Fink, 1989; Smarzynska Javorcik, 2004b). Park and Lippoldt (2005) as well as Nicholson (2007) come to the conclusion that at certain high levels of IPR protection, FDI may well be substituted by licensing. The latter group of results are again explicable by the control issue: "risk-reducing consid-erations may push firms that have proprietary products or technology to choose higher control modes" (Agarwal and Ramaswami, 1992, p. 8). This is further developed in Eaton and Gersovitz (1983), where higher control modes are assumed to allow firms to selectively transfer technology so as to make this technology less profitable in case of a leakage or in case of expropriation of foreign assets[6]. Meyer (2001) suggests that this vari-ance in results may best be disentangled when distinguishing between the transaction costs of establishing a wholly-owned foreign affiliate and the costs of operating a subsidiary with a lower ownership share (ibid., p. 364). Teece (1986) suggests that ownership vs contractual arrangements in weak IPR regimes may well depend on the importance of complemen-tary assets, whereby 'bottlenecks' (for example distribution and special-ised manufacturing competences) would in general suggest ownership as the preferred solution.

The literature that offers an application or test of the above insights on institutional determinants of FDI decisions and entry modes for the case of CEECs is reviewed comprehensively by Meyer and Peng (2005) with a view on, amongst other things, foreign investors' entry strategies in CEECs: Bevan et al. (2004) establish that it is in particular the set of formal, as against informal, institutions that significantly influences the inflow of FDI into CEECs. A positive influence could be established for the share of private ownership of business (resulting from privatisation), the depths and comprehensiveness of banking sector reforms, liberalisation of foreign exchange and trade, and legal developments[7]. Peng (2003) hypothesises that in the absence of formal market-supporting institutions and where, hence, informal institutions prevail as constraints in CEECs, foreign investors will prefer joint ventures and other contractual arrangements over wholly-owned foreign investments, and Meyer (2001) establishes that with success in institution building, wholly-owned foreign investments will prevail. Meyer (2001) also suggests that the transfer of technology into CEECs is more likely for wholly-owned foreign investment projects as compared to a mode of internationalisation via foreign trade. In terms of transaction costs, these results suggest that the costs of establishment (which in CEECs typically included negotiations with privatisation agencies) may have been "a stronger deterrent than the potentially high coordination costs in JVs" (Meyer and Peng, 2005, p. 604). Or in a strategic management perspective: foreign investors will choose to play the "networking game" that substitutes for weak market-supporting institutions (Peng, 2003, p. 286).

7.1.3   IPR regime and technology transfer

Another part of the literature on FDI and IPR assesses the effects of IPR regimes on direct technology transfer. A comprehensive review of the literature (Arora, 2009, pp. 47–49) suggests that direct technology transfer tends to increase with improvements in IPR regimes, patent protection in particular. Some qualifications of this result can be found, yet the general result of a positive relationship holds, even if not always significant. In particular, Eaton and Kortum (1996) analyse productivity growth and technology diffusion in the OECD countries (all of which can be assumed to have rather stringent IPR regimes), and find that the smaller and less technologically advanced OECD countries derived most of their productivity growth from having foreign inventors filing patents in their economies, a clear sign of direct technology transfer. Branstetter et al. (2005) find that if a gap in IPR protection between home and host locations exists, then the amount of technology transferred from parent to foreign affiliate will tend to increase with the strength of IPR protection in the hosting country. Smith (1999) measures the inflow of knowledge (as R&D expenditures undertaken by the foreign affiliate on behalf of foreign investor) and is able to show that this is strongly and positively affected by patent rights in the host economy (this, however, applies only to recipient countries with strong imitative abilities). McCalman (2001) finds that a strengthening of patent protection leads to an increase in the contribution of patents to economic growth. Xu and Eric (2005) show that the extent to which countries source technology from abroad, and the way in which they do it, depend on their level of economic development: rich countries benefit both from domestic technology and from foreign technology that is embodied in imported capital goods. Middle-income countries also benefit from embodied technology in imports and additionally from access to external technology by way of foreign investors filing patents in the host country. Developing countries, due to a lack in absorptive capabilities and a comparatively weak IPR regime, rely mainly on foreign patents. Like Smith (1999), Bascavusoglu and Zuniga (2002) find that patent protection seems to matter most for countries with strong imitative abilities (ample supply of engineers and scientists, and their own R&D) and additionally for industries with a medium level of R&D intensity (see the discussion of the two faces of R&D initiated by Cohen and Levinthal, 1989). Belderbos et al. 2006 find that developing countries' efforts to strengthen IPR protection regimes helps them attract more R&D activity by foreign multinationals. Finally, Smith (2001) finds significant evidence that stronger patent protection increases sales and licensing payments by foreign affiliates and, as in his previous contribution, the result is driven by countries with strong imitative capacities.

These insights suggest consideration of the ease with which intellectual property rights can be specified when analysing the relationship between IPR and FDI. Where the foreign investment contributes to the development of new, or modifies existing, knowledge and technology of the foreign investor network, the assignment of intellectual property rights becomes problematic and appropriability hazards will tend to emerge (Oxley, 1999, pp. 285–6). This is due to the uncertain nature of inventions and to the complex, tacit character of some of the resultant technological know-how (see for example Teece, 1986). Only where existing knowledge and technology are used and remain unaltered does the focus shift to the ability of the IPR regime to enforce existing intellectual property rights.

A positive relationship is also established between the level of protection of the IPR regime and the quality of knowledge and technology transferred in internationalisation projects (for example Rockett, 1990;

Park and Lippoldt, 2005). Oxley (1999) likewise suggests that firms are reluctant to transfer 'advanced' technology to contract partners, if no ownership is involved and if the IPR regime in the host country is weak (p. 287). The results on both the amount and the quality of technology transfer are plausible: investors that have a knowledge advantage over their rivals will be deterred from investing and transferring sensitive knowledge into host countries, where protection of their knowledge-based assets is either too expensive or not possible at all (the "appropriation hazard" of Oxley, 1999).

Pertaining to transition economies in Eastern Europe, case studies conducted by Sharp and Barz (1997) conclude for the chemical and pharmaceutical industries that companies pay close attention to the risk of piracy due to weak IPR protection, and are therefore sceptical about transferring technology to these countries.

7.1.4 IPR regime and industrial branches

Because sectors differ in the ability to generate product and process innovations and to copy knowledge and technology from others (imitation), the IPR issue is typically treated as an **industrial branch-specific** problem: chemical or pharmaceutical industries, for example, depend more heavily on IPR protection, whereas sectors like manufacturing of basic metals or food products are rather less affected by the quality of the IPR regime in the host economy. The distinction between IPR-sensitive and IPR-insensitive sectors was first described by Mansfield (1986, later revisited in 1995) and has ever since been widely cited and used by other researchers in papers on this issue (see for example Lee and Mansfield, 1996; Maskus, 2000, table 1, p. 6; Nunnenkamp and Spatz, 2003; Ostergard, 2000; Smarzynska Javorcik, 2004b; and UNCTAD, 1993). Extending the list of branches, Band and Katoh (1995) and Ebanks (1989) find that the software and entertainment industries particularly depend on IPR protection via patent and copyright (alas, those sectors are not included in the IWH FDI Micro Database). Possibly, the 1986 study could not identify those two sectors as IPR-sensitive, because technology in these fields has evolved in great leaps ever since that time. The issue is also treated as specific to the kind of technology involved: standardised, labour-intensive production depends on IPRs to a lesser degree than do complex but easily copied product technologies such as in pharmaceuticals (see for example Maskus, 2000). In the Smarzynska Javorcik (2004b) study, consideration of the industrial-sectors issue leads to the result that a weak IPR regime shifts the focus of FDI projects from manufacturing to distribution (p. 40). A further implication of the sectoral distinction is that the relevance for IPR protection in the host economy will depend on whether technologies had been transferred with an intention to serve the domestic (recipient) market or whether for exports from the host economy to other, third-party, countries: in the case of the latter, and if third-party countries do have stricter enforcement of IPR rules, the IPR regime of the technology-receiving (host) country may not even play much of a role. Here, the owner of the knowledge (the originator) is able to protect his knowledge by having exports that infringe his rights blocked by third-party countries (Arora, 2009, pp. 44–45).

7.1.5 IPR regime and foreign affiliate business functions

Perhaps even more importantly, the decision about the extent and character of internationalisation is treated as specific to the type of business functions that the foreign affiliate is entrusted with: here, distinction is made between sales and distribution outlets (which depend the least on IPR regimes), rudimentary production and assembly, manufacture of components or complete products, and finally R&D, which depends the most on strong IPR protection (for example Mansfield, 1993, p. 112; Mansfield, 1995). With respect to R&D, Nunnenkamp and Spatz (2003) find that the strengthening of the IPR regime not only affects the amount of inward FDI but also the amount of R&D expenditure by foreign subsidiaries (this is complemented by an increase in value added and exports created by subsidiaries, ibid., p. 39). Lee and Mansfield (1996) hold for the chemical industry that foreign affiliates in countries with strong IPR regimes tend to conduct more R&D and have a higher production than their counterparts in weaker IPR systems. Maskus (1998) comes to the conclusion that a strengthening of an IPR regime in a country is followed by an increase in the number of patent applications by foreign affiliates and an increase in affiliate sales of patented products. Nunnenkamp and Spatz (2003) find that a shift to a stronger IPR system raises the amount of exports and R&D by FDI affiliates. Interestingly, some foreign investments often conduct R&D even if their host economy IPR regime is considered weak (Zhao, 2004). Examining this conundrum, Zhao coins the expression of "internalization-arbitrage" (p. 2) by which MNEs locate only a particular part of R&D in less-protected host economies. This part is either not easily copied or provides little value without complementary R&D and technology (which is then located in better-protected locations of the MNE). Some of this R&D may be targeted at increasing absorptive capabilities at the technology-receiving end. Additional explanations for this conundrum may be found in Levin et al. (1987), and include mechanisms like the exploitation of lead time or head start moving rapidly down the learning curve, the use of complementary sales and service capabilities, and finally outright secrecy[8], to exploit technology and knowledge generated in locations with weak protection.

In sum, the relationship between IPR and FDI is quite complex and depends on, or influences, many factors, such as those described above: FDI decisions and entry modes (corporate governance), amount and quality of technology transferred, the way that technology and knowledge transferred to the foreign investment is used, industrial-branch specificity, type or strategic motive of investment, and business functions of the foreign affiliate. With respect to the main focus of interest here, the strength of the IPR regime, some general conclusions, however, may be drawn: "[o]n the one hand, a weak IPR regime increases the probability of imitation, which makes a host country a less attractive location for foreign investors. On the other hand, strong protection may shift the preference of multinational corporations from FDI towards licensing." (Smarzynska Javorcik, 2004b, p. 40).

In general, this review concludes with Alcácer and Chung (2007) that research on the role of IPR regimes has typically treated FDI strategies as rather passive (with the notable exemption of Jandhyala, 2008; Faria and Sofka, 2008). The following analysis attempts to push the research envelope in this respect. Furthermore, this analysis considers both the institutional environment with respect to IPR regimes and the mechanisms of governance of FDI projects, as proposed by Oxley (1999), to generate a more "complete understanding of the organization of inter-firm alliances" (ibid., p. 285).

7.2 The data and methods used in the analysis

The relevance of some of the issues identified above for the case of CEECs can be tested by use of the IWH FDI Micro Database of 2006–2007.

As a proxy for the strength of the IPR regime, the country-specific indicator of the 'Protection of intellectual property' section of the 2005/2006 Global Competitiveness Report of the World Economic Forum (WEF) is used. This indicator is generated by questionnaires and reflects the opinion of relevant actors. The indicator is defined as discrete variables along an ordinal Likert scale, ranging from 7.0, which would equal the world's most stringent intellectual property protection, to 1 for weak or non-existent protection. The scores clearly rate the countries in CEE lower than the typical countries where foreign investors in CEE originate: Germany ranks first with 6.6. Finland scores 6.4, UK 6.2, France and

Japan 5.9, USA 5.7, and Austria 5.6. Amongst the post-transition economies rated, we find: 3.1 for Romania, 3.6 for Poland, 3.7 for Croatia, and 4.5 for Slovenia. Because of the high correlation between the level of economic development and the strength of the respective country's IPR regime, this indicator also controls for other country-specific unobserved factors. In fact, the analysis does not only consider the quality of the host country's IPR regime but also includes a comparison between the IPR regimes of the host country and the home country of the foreign investor, what is called here the IPR gap (calculated as WEF indicator for home minus WEF indicator for host). After all, the role of IPR on FDI can be expected to depend on the relative strength of IPR regimes and not on absolute values of a particular country.

The magnitudes of IPR gaps between foreign investors from all around the world and the host economies in CEE assumes a positive value where the IPR regime in CEECs offer less protection than in the home country of the foreign investor, and a negative value in the opposite case. With most foreign investors into CEECs originating from Western Europe (nearly 80 per cent of FDI into the 10 new EU member states originate from the 15 old EU member states, see Hunya, 2010, p. 41), gaps should be expected to tend to be positive (even if some foreign affiliates experience negative gaps *vis-à-vis* the home countries of their investors, those amount to an accumulated 26 per cent) (see Figure 7.2).

From the insights generated from the literature sketched above, a number of hypotheses that are testable with a firm-level dataset of foreign affiliates in CEECs and East Germany have been developed. The hypotheses evolve around three central questions: (i) on the role of the IPR regimes or gaps on foreign affiliates' technological activity, (ii) on technological embeddedness in their respective host economies, and (iii) on the question of how foreign investors control their affiliates in comparatively weaker IPR regimes (the corporate governance issue). To test these hypotheses, the sizes of IPR gaps are related to a variety of firm-specific determinants of technological activity in R&D and innovation, internal technology transfer, embeddedness into the local innovation system of the host economy via customers, suppliers and science/research institutions, and finally the two instruments of corporate control of the size of equity share and level of autonomy in key business functions.

The first of those indicators is the one on internal product-embodied technology transfer already used in Chapter 6 (*tt.mnc*). The variable is operationalised as a discrete variable along an ordinal Likert scale that ranges from 1 for embodied technology received from the foreign investor network being "not important" to 5 for "extremely important".

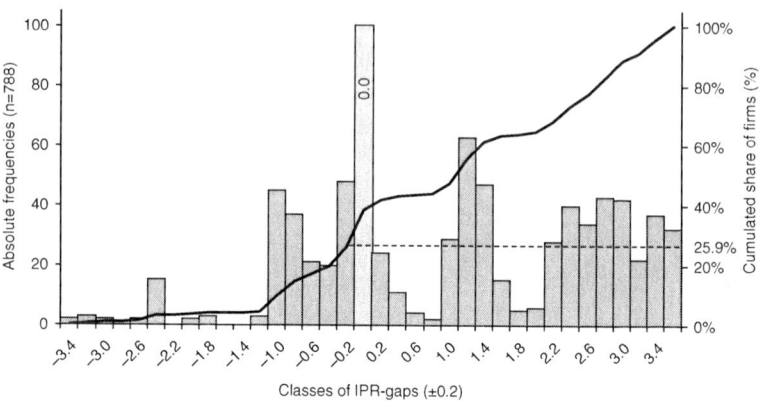

Classes of IPR-gaps (±0.2)

Figure 7.2 Observed IPR gaps between host and home countries (*IPR.gap*) of the IWH FDI Micro Database of 2006–2007

Data sources: WEF (2007), Section IX, and IWH FDI Micro Database of 2006–2007.

The second set of two indicators pertains to foreign affiliates' own technological activity and is used in the preceding Chapters 5 and 6: the share of expenditure for R&D in per cent of total sales (now abbreviated: *sh.R&D*), and the share of new or significantly improved products in total sales (*sal.inno*). Both indicators are continuous variables that range from 0 to 100 per cent.

Technological embeddedness is proxied by six indicators: the first set of indicators measures (i) the level of importance of foreign affiliates' own technological activity as a source of knowledge and technology (*own.tech*), (ii) the level of relative importance of R&D and innovation by the foreign investor's headquarter in comparison to *own.tech (comp. HQ)*, and (iii) the comparative level of importance of technological activity by other affiliates of the foreign investor in comparison to *own. tech (comp.FA)*. Comparisons are made between the levels of importance of foreign investor headquarters to own technological activity and other foreign affiliates to own technological activity by comparing values for each source at the firm level: they are constructed by way of subtracting the levels of importance of foreign investor's headquarters technological activity (for ii) from the levels of importance of own foreign affiliate technological activity (*own.tech*), and by subtracting the levels of importance of other foreign affiliates (for iii) from the levels of importance of own foreign affiliate technological activity (*own.tech*). Those two comparative indicators reflect Cantwell's internal network (see Cantwell and Piscitello, 2002). All original indicators are also discrete variables along an ordinal Likert scale ranging from 1 to 5 with increasing levels of importance, the comparative indicators consequently assume values between -4 and +4. The comparative variables are defined to increase with the importance of own technological activity *vis-à-vis* the internal network.

Technological embeddedness within the host economy is measured by a comparison between the importance of local vs foreign scientific institutions as a source of knowledge and technology for the foreign affiliate's R&D and innovation (*comp.SI*), and by the share of sales or supplies that are transacted with other partners of the foreign investor network as against the resulting shares transacted with domestic buyers (*comp.sal*) and domestic suppliers (*comp.sup*). The comparative indicator pertaining to scientific institutions rises with the relative importance of local scientific institutions *vis-à-vis* foreign institutions, and again assumes values between -4 and +4. The two indicators for the trade relationships are continuous variables that range from 0 to 100 per cent. The indicators *comp.sal* and *comp.sup* rise with increased relative trade relationships of foreign affiliates with the internal network of the foreign investor, hence falling local embeddedness in terms of trade (note that these indicators are different from those in Chapter 5, where *loc.sales* and *loc.suppl* denoted the share of sales and procurement from the host economy).

To control for other factors that may influence the relationship between IPR regimes and firm-specific determinants of technological activity and embeddedness into the local innovation system (that is, firm and industry heterogeneity), a number of additional variables are added: (i) firm-specific information on the importance of R&D carried out in collaboration with scientific institutions that are not located in the host economy (*for.SI*) and those that are located in the host economy (*dom.SI*), both on a Likert scale ranging from 1 for "not important" to 2 for "little importance", 3 for "important", 4 for "very important", and 5 for "extremely important"; (ii) the level of importance of the following strategic motives pursued by the foreign investor (with the possibility of multiple motives[9]): market access on the domestic market (*d_market*), follow foreign key clients (*d_client*), increase efficiency across the foreign owner network (*d_effic*), access location-bound natural resources (*d_nat. res*), and access location-bound knowledge, skills, and technology (*d_know*); (iii) the age of the foreign investment project (years since the entry of the foreign investor: *age*); (iv) the size of the foreign affiliate in terms of total number of employees (*size*); (v) a dummy for whether the foreign investment project is a greenfield investment, identified if the initial mode of entry of the foreign investor was a partial or full ownership in or of a completely new enterprise (*d_green*); and (vi) a set of 12 industry dummies at the NACE two-digit level (*d_da* through *d_dn*).

Because the data used in the analysis may not always be precise, as it reflects opinions (the subjectivity-issue), and because of the varying definitions and dimensions of the indicators, the empirical analysis uses a variety of methods, ranging from bivariate and rank correlation analyses to partial rank correlations, ordered probit, and linear OLS regression models. The choice of methods, however, remains consistent throughout the various steps of analysis in the set of determinants. To account for possible asymmetries in the data and non-linearities in relationships, the data is transformed into natural logs for the OLS regression analyses. Here, the analysis additionally corrects for possible heteroskedasticity by using White robust standard errors (heteroskedasticity-robust standard errors).

The partial rank correlation analysis measures the degree of association between two variables, randomly selected from the list of factors, and removes the effect of the rest of the factors as controlling variables by holding them constant. This allows control for effects of third variables that may have an influence on the two variables of a pairwise correlation:

$$\tau_{(X,Y)/U} = \frac{\tau_{X,Y} - \tau_{X,U} \cdot \tau_{Y,U}}{\sqrt{\left(1 - \tau_{X,U}^2\right)\left(1 - \tau_{Y,U}^2\right)}}$$

with X being the sensitivity of FDI subsidiaries to the strength of the IPR regime in the host economy, and Y the randomly selected variable from the list of factors. U denotes the rest of the factors that are used as controlling variables. The partial correlation tests linear regressions between X and U as well as between Y and U. The correlation coefficient $\tau_{(X,Y)/U}$ is the correlation between the residuals of the linear regressions.

Before the actual (partial) correlation and regression analyses are conducted, the indicators are first tested for significant and high rank–pairwise correlations (see Table 7.1). This is to ensure that information contained within one factor is not significantly duplicated by another. In fact, high significant correlations only occur for variables that are not used simultaneously in empirical analysis: *own.tech* and *comp.FA* (0.67), *comp.FA* and *comp.HQ* (0.68) are obviously related, and hence alternatives in the analysis of hypothesis H(3) (see also the annex to this chapter); *comp.HQ* and *HQ.tech* (-0.65) are also correlated, which is plausible; *comp.HQ* and *autono* (0.53) is also plausible (own technological activity is higher than technology carried out at the headquarters, if autonomy is large); *comp.SI* and *dom.SI* (0.63) and *comp.sal* and *sal.HQ* (0.89) and *comp.sal* and *sal.dom* (-0.67), *comp.sup* and *sup.HQ* (0.84), and *comp.sup* and *sup.dom* (-0.57) are tautological; the correlation between *for.si* and *dom.si* (0.57) shows that technological cooperation with domestic or foreign scientific institutions may be complementary, as is the importance of R&D carried out at the headquarters (*HQ.tech*) and by other foreign affiliates of the investor (*FA.tech*) (0.53). The high and significant correlation between embodied internal technology transfer (*tt.mnc*) and sales to domestic customers (*sal.dom*) (-0.62) may be spurious and hence warrant their simultaneous inclusion into empirical analysis, as they contain independent information.

7.3 Hypotheses development and empirical testing

7.3.1 IPR gaps and direct technology transfer

With respect to the role of IPR gaps between home and host countries on direct technology transfer, the fairly unambiguous picture derived from the literature is tested, that is, that the amount of technology and knowledge transferred from parent to foreign affiliate will increase with the quality of the IPR regime in the host country.

H(1) The larger the IPR gap between the foreign investor's home country and the country hosting the foreign affiliates, the less will the foreign affiliate receive technology directly from its parent.

This hypothesis is tested by assessing the relationship between the size of the IPR gap and the indicator of internal product-embodied technology transfer already used in Chapter 6 (*tt.mnc*). For the analysis, Spearman rank correlation analysis is used, because *tt.mnc* is defined on an ordinal scale. As a further analytical step, the same relationship is tested by way of a partial rank correlation analysis; this is to account for the possibility that the bivariate correlations may be spurious, omitting third variables. The additional indicators include the firm-specific determinants listed above (technological activity, embeddedness into the local innovation system of the host economy, and instruments of corporate control) and help to account for heterogeneity. The data is ranked for the partial correlation analysis to account for the ordinal scale of *tt.mnc*. Finally, the hypothesis is tested by use of an ordered probit analysis[10] including all other variables of the partial correlation analysis.

The results of all three tests turn out to be positive (see Table 7.2): the larger the IPR gap, the more will the foreign affiliate receive already

Table 7.1   Pairwise rank correlation between all determinants

	tt. mnc	sh. R&D	sal. inno	own. tech	comp. HQ	comp. FA	comp. SI	comp. sal	comp. sup	for. si
sh.R&D	.06									
sal.inno	.15*	.40*								
own.tech	.20*	.40*	.39*							
comp.HQ	−.31*	.21*	.21*	.58*						
comp.FA	−.12*	.29*	.20*	.67*	.68*					
comp.SI	.05	.12*	.21*	.16*	.12*	.12*				
comp.sal	.15*	.02	.02	−.09*	−.23*	−.17*	−.05			
comp.sup	.24*	−.04	.03	−.11*	−.36*	−.20*	−.13*	.41*		
for.si	.27*	.14*	.15*	.30*	−.07	−.03	−.20*	.05	.07	
dom.si	.26*	.22*	.30*	.39*	.07	.08	.63*	.00	−.07	.57*
HQ.tech	.57*	.11*	.10*	.21*	−.65*	−.19*	−.00	.20*	.33*	.37*
FA.tech	.40*	.11*	.20*	.28*	−.22*	−.47*	.08*	.12*	.13*	.42*
sal.HQ	.16*	−.02	.02	−.10*	−.25*	−.18*	−.06	.89*	.36*	.07
sal.exp	−.13*	.06	.04	.17*	.30*	.20*	.04	−.41*	−.32*	.02
sal.dom	.03	−.05	.06	.08	.05	.07	.16*	−.67*	−.18*	−.03
sup.HQ	.20*	−.09*	−.01	−.14*	−.39*	−.22*	−.15*	.42*	.84*	.04
sup.imp	−.06	.03	.02	.06	.18*	.10*	.01	−.12*	−.38*	.00
sup.dom	−.08*	.06	.05	.14*	.18*	.12*	.18*	−.24*	−.57*	−.00
autono3	−.21*	.18*	.13*	.25*	.53*	.34*	.06	−.27*	−.36*	−.02
equity.sh	.07	−.01	−.02	−.04	−.17*	−.06	−.01	.19*	.17*	−.08*
age	.00	−.03	−.05	.03	−.01	.07	.03	−.05	−.05	−.07
size	.10*	−.03	−.02	−.03	−.08	−.05	−.04	.15*	.07	.08*

Note: All pairwise correlations were conducted as Spearman rank correlations using casewise deletion, where observations are ignored if any of the variables in the list of variables are missing: obs=461. Significant correlations (at a level of .1) are marked with an asterisk.

Source: IWH FDI Micro Database 2006–2007.

dom. si	HQ. tech	FA. tech	sal. HQ	sal. exp	sal. dom	sup. HQ	sup. imp	sup. dom	auto no3	equity. sh	age	si size
.28*												
.41*	.52*											
−.01	.21*	.13*										
.03	−.20*	−.06	−.30*									
.09*	.00	.01	−.62*	−.16*								
−.10*	.32*	.12*	.46*	−.25*	−.23*							
.02	−.16*	−.04	−.06	.35*	−.13*	−.26*						
.15*	−.09*	.04	−.23*	.07	.34*	−.48*	−.33*					
.06	−.41*	−.15*	−.27*	.21*	.10*	−.36*	.16*	.18*				
−.07	.18*	.05	.17*	−.16*	−.10*	.17*	−.06	−.12*	−.23*			
−.04	.04	−.06	−.06	−.01	.08*	−.08*	−.02	.09*	−.09*	.10*		
.03	.07	.06	.20*	.07	−.21*	.10*	.10*	−.08*	−.05	.08*		.10*

Table 7.2　IPR gaps and direct technology transfer embodied in rather standardised products

	obs.	Spearman's rho	P >|t|
tt.mnc	685	0.23	0.000
	obs.	Partial correlation	significance
tt.mnc	534	0.09	0.034

	obs.	ordered probit coefficient	P >|z|	Pseudo R^2
tt.mnc	522	0.10	0.002	0.189

Note: The partial correlation and ordered probit results for the other determinants included in the analysis are presented in the Appendix to this chapter, Table A1.

existing and product-embodied technology of the foreign investor. This counterintuitive result may be rooted in the particular specification of internal technology transfer: the kind of technology received in *tt.mnc* is in no need of protection, because any competitor can legally acquire it and can use it without substantial adjustments (it is embodied in products). What is measured here is the fraction of internal technology transfer that is not sensitive to IPR protection, and the positive association suggests that if technology is transferred, then it will only be technology that is not in need of protection from unwarranted use by competitors.

7.3.2　IPR gaps and foreign affiliate technological activity

The following analysis, again testing internal technology transfer, is aligned with the analysis of Smith (1999) on foreign affiliate R&D expenditure and patent rights. R&D and innovation expenditure is used to proxy technological activity of foreign affiliates in CEECs, and this is related to the IPR gap.

> H(2)　Foreign affiliates that operate under a positive IPR gap between host and home countries will be technologically less active.

In the literature (see in particular Bascavusoglu and Zuniga, 2002; Smith, 2001), the positive relationship between IPR regime and technology transfer was often only significant for countries with strong imitative capabilities (qualified personnel and own R&D). Firstly, controlling for such capabilities would add variables that might be highly correlated with R&D and innovation (heteroskedasticity). Secondly, it may be safely assumed that all CEECs are better able to absorb foreign technology than the developing countries that gave rise to the qualification found in the empirical literature.

When following Teece (1986) and Oxley (1999), the hypothesis becomes less straightforward: where foreign affiliates participate in technological activity for and/or in collaboration with the foreign investor network, the strength of protection of IPRs in the host economy may play a lesser role due both to the uncertain and tacit nature of the knowledge generated and to the assumed difficulties with attaching claims to the intellectual property created. Only where foreign affiliates do not augment or alter the technology they receive will the IPR regime play an important role. This qualification, however, should not change the assumed positive relationship between the size of the IPR gap and affiliate technological activity; it may make it weaker.

The hypothesis is tested by operationalising foreign-affiliate technological activity by two of the proxies used in Chapters 5 and 6: the share of expenditure for R&D in per cent of total sales (now abbreviated: *sh.R&D*), and the share of new or significantly improved products in total sales (*sal.inno*). Spearman correlation analyses between the IPR gap and either the share of R&D in sales, or the share of innovations in sales, or both, turn out to be negative and highly significant, as expected. Even when the analysis controls for the set of other firm-specific determinants by way of partial correlation analyses, the results stay qualitatively the same (see Table 7.3). The analysis hence finds unambiguous empirical support for hypothesis H(2) which suggests that in fact foreign affiliates that operate under an IPR gap between home and host countries are technologically less active. This relationship was deliberately not tested by use of a regression analysis, because this would impose

Table 7.3 IPR gaps and foreign affiliate technological activity

	obs.	Spearman's rho	P >|t|
sh.R&D	645	−0.23	0.000
sal.inno	719	−0.14	0.000
	obs.	Partial correlation	significance
sh.R&D	453	−0.09	0.073
sal.inno	500	−0.11	0.021

Note: The partial correlation results for the other determinants included in the analysis are presented in the Appendix, Table A2.

the assumption that foreign-affiliate technological activity is largely determined by technological embeddedness and corporate control only, which is clearly insufficient.

7.3.3　IPR gaps and sources of technological knowledge for the foreign affiliate

According to the literature, the amount and quality of direct technology transferred to FDI subsidiaries positively depends on the strength of the protection of firm-owned knowledge by the IPR regime (see for example Rockett, 1990; Park and Lippoldt, 2005; Branstetter et al., 2005). One of the important particularities of TNCs is that they are able to source resources such as knowledge and technology from all locations that the network is engaged in (locational diversification of technological activity). Cantwell identifies, in his concept of technological accumulation, two groups of sources of knowledge and technology that foreign affiliates may use: the foreign investor network and the external networks of the foreign affiliate in its host location (see for example Cantwell and Piscitello, 2002). It is obvious that the IPR regime will play a role in determining the (relative) importance of either source of knowledge and technology for the foreign affiliate.

H(3) Where foreign affiliates that operate under a positive IPR gap do own R&D and innovation activities, this technological activity will be less important as a source of technological knowledge for the foreign affiliate than knowledge and technology sourced from the foreign parent network.

In a first step, the analysis tests whether foreign affiliates hold that for their technological activities, own R&D (*own.tech*) becomes less with rising IPR gaps. The assumption behind that is that R&D in weaker IPR regimes is viewed as less preferable in any case. Because the question of intra-TNC network sources of knowledge and technology may be an issue of comparative levels of importance, this hypothesis is further tested by comparing the importance of own technological activity (*own. tech*) with (i) knowledge generated by headquarter R&D (*comp.HQ*) (that is, home-base exploiting) and knowledge generated by other foreign affiliates of the foreign investor's network (*comp.FA*). The two comparative variables correspond to the internal network in Cantwell terms.

The empirical methods used here again include Spearman, partial rank correlations, and ordered probit regression analyses. All tests lend support to hypothesis H(3) (see Table 7.4) and suggest that with increasing

Table 7.4 IPR gaps and importance of sources of knowledge and technology for foreign affiliate technological activity

	obs.	Spearman's rho	P >|t|
own.tech	704	−0.13	0.000
comp.HQ	687	−0.26	0.000
comp.FA	660	−0.23	0.000
	obs.	Partial correlation	significance
own.tech	534	−0.12	0.006
comp.HQ	534	−0.16	0.000
comp.FA	534	−0.11	0.016

	obs.	ordered probit coefficient	P >|z|	Pseudo R^2
own.tech	522	−0.13	0.000	0.144
comp.HQ	522	−0.12	0.000	0.144
comp.FA	522	−0.08	0.007	0.067

Note: The partial correlation and ordered probit results for the other determinants included in the analysis are presented in the Appendix, Table A3.

IPR gaps foreign affiliates' own technological activity in a country with less protection becomes less important *per se*, and in particular also less important than the sources of knowledge and technology that lie within the internal network of the foreign investor and hence are under better control by the TNC. This also applies to own R&D as a source for innovation in the foreign affiliate. However, pseudo R^2 are rather low for all three ordered probit tests. Nevertheless, they unambiguously suggest the same interpretation.

7.3.4 IPR gaps and technological embeddedness in the local host economy

With respect to the technological role of Cantwell's external networks, it should be expected that the level of protection granted by an IPR regime is relevant: here, Oxley's 'appropriability hazards' work directly. A large IPR gap should hinder the intense technological embeddedness of a foreign affiliate. The first local-embeddedness indicator pertains to the relative importance of local vs foreign scientific institutions as a source of knowledge and technology for the foreign affiliate's R&D and innovation (*comp.SI*).

The two local-embeddedness hypotheses are:

H(4) Foreign affiliates that operate under a positive IPR gap between host and home countries will make less intense use of the domestic economy for its supplies and its sales.

and

H(5) Where foreign affiliates that operate under a positive IPR gap do own R&D and innovation activities, the knowledge and technology they can acquire from local science institutions will be of less importance than those from foreign science institutions.

The tests of hypothesis H(4) are performed by use of Spearman correlations between the size of the IPR gap and *comp.sal* or supplies *comp.sup*, by partial rank correlations, and OLS regression analysis[11]. Hypothesis H(5) is tested with Spearman correlations between the size of the IPR gap and *comp.SI*, by partial rank correlations, and by an ordered probit regression due to the dimension of the variable.

The results for the tests pertaining to customers and scientific institutions are unambiguous in all specifications, and suggest that the larger the IPR gap the more will foreign affiliates engage with internal network customers (positive sign), and the less will they engage with local scientific institutions than foreign scientific institutions (negative relationship) (see Table 7.5). This, hence, lends some support to hypothesis H(4), even if not for suppliers: here the results remain insignificant. With respect to the role of the location of scientific institutions as sources of knowledge and technology for a foreign affiliate's own technological activity, the tests find unambiguous results in support of hypothesis H(5). This suggests that with increasing IPR gaps, foreign affiliates are less likely to source knowledge and technology from local scientific institutions as major players in host innovation systems, but rather rely on scientific institutions in other countries (presumably in countries with stricter IPR regimes, but not necessarily so).

In sum, the tests suggest that the higher the IPR gap, the less will foreign affiliates be technologically embedded into their host economy, and the less likely they are to engage in the host innovation system. It may be assumed that 'appropriability hazards' contribute to this: sourcing and selling in local industries may contain the threat that product or service specifications carry sensitive knowledge and technology, the intellectual property of which would be difficult to commercialise in an

Table 7.5 IPR gaps and technological embeddedness in the local host economy

	obs.	Spearman's rho		P >|t|
comp.sal	760	0.18		0.000
comp.sup	752	0.16		0.000
comp.SI	674	−0.12		0.002
	obs.	Partial correlation		significance
comp.sal	534	0.09		0.048
comp.sup	534	0.06		0.179
comp.SI	534	−0.08		0.061
	obs.	linear OLS regression	P >|t|	R^2
comp.sal	522	2.36	0.014	0.384
comp.sup	522	0.53	0.573	0.271
	obs.	OLS in natural logs	P >|t|	R^2
comp.sal	522	1.49	0.034	0.381
comp.sup	522	0.25	0.724	0.288
	obs.	ordered probit coefficient	P >|z|	Pseudo R^2
comp.SI	522	−0.06	0.089	0.050

Note: The partial correlation and ordered probit results for the other determinants included in the analysis are presented in the Appendix, Table A4.

environment of weak IPRs. This means that host economies with weak IPR protection are likely to forgo the kind of technological effect that inward FDI may otherwise have.

7.4 Summary of main results

The results of the analyses clearly show that weak IPR regimes in CEECs restrict the benign technology effects commonly associated with inward FDI. In particular, where IPR regimes offer only little protection, the kind of technology that foreign affiliates receive from parents appear to be rather product-embodied and standardised, that is, less sensitive to protection. Moreover, where foreign affiliates operate in environments with a positive IPR gap, the less sensitive kind of direct technology transfer also prevails where autonomy over business functions on behalf of foreign affiliates *vis-à-vis* their parents is rather high. These results help to disentangle the ambiguous picture drawn from the relevant literature: when controlling for heterogeneities, as in particular the size of the equity share, the relationship between the quality of the IPR regime and FDI and technology transfer becomes clearly positive.

Further, foreign affiliates that operate in a host economy with a positive IPR gap are technologically less active in terms of spending for R&D and generating innovations. Yet R&D increases with the degree of autonomy of foreign affiliates from their parents. Where foreign affiliates are nevertheless technologically active, the sources of knowledge and technology for their own technological activity rest increasingly within the internal network of the foreign investor. This contributes to an affirmative answer to the first research question of this chapter (RQ7_1): the comparatively weaker IPR regimes in CEECs do in fact negatively affect the technological activity of foreign affiliates.

Where foreign affiliates in the weaker IPR countries of CEECs still receive knowledge and technology from their foreign investors (even if somewhat product-embodied and standardised) and/or nevertheless generate new knowledge and technology by way of R&D and innovative activities, there is the danger that knowledge may unintentionally dissipate to other players in the host economy and thereby reduce the ownership advantage of the foreign investment. The results show that the less technologically embedded foreign investments tend to be in their host economies, the larger the IPR gap to the foreign investors' home countries. This is all the more the case where the foreign affiliate enjoys large extents of autonomy from parents and where equity shares held by the foreign parent are comparatively low.

The embeddedness issue can be an instrument for the foreign investor to prevent unintended spillovers of sensitive knowledge and technology: collaborative R&D with domestic science institutions may give rise to new knowledge whose intellectual property may be difficult to retain. This is particularly so where either the formal institutional environment to protect this is not in place or where the informal institutional framework, that is the enforcement of a legally secured right, is not possible or involves prohibitive transaction costs. Quite clearly, foreign affiliates in CEECs with a sizeable IPR gap tend to collaborate more intensively with foreign scientific institutions. The same could be established for technological embeddedness with local customers, the importance of which as a technology source (demand-driven innovation) diminishes with rising IPR gaps. The second research question of this chapter (RQ7_2) can likewise be answered in a positive way: the comparatively weaker IPR regimes in CEECs do inhibit the technological embeddedness of foreign affiliates into their host innovation systems.

7.5 Annex (complete econometric test results)

Table A1 IPR gaps and direct technology transfer embodied in rather standardised products: partial correlation and ordered probit results with *tt.mnc*

Variable	Partial correlation		Ordered probit analysis			
	Corr.	Sig.	Coef.	Robust Std. Err.	z	P>|z|
IPR.gap	0.0945	0.034	0.1028	0.0331	3.11	0.002
own.tech	0.1163	0.009	0.1555	0.0521	2.98	0.003
HQ.tech	0.3811	0.000	0.3971	0.0606	6.55	0.000
FA.tech	0.1096	0.014	0.1390	0.0568	2.45	0.014
for.SI	−0.0698	0.119	−0.1517	0.0725	−2.09	0.037
dom.SI	0.0776	0.083	0.1302	0.0543	2.40	0.017
sal.HQ	0.0428	0.339	−0.0055	0.0041	−1.34	0.179
sal.exp	0.0229	0.609	−0.0055	0.0039	−1.41	0.160
sal.dom	0.0091	0.840	−0.0061	0.0037	−1.66	0.097
sup.HQ	−0.0194	0.665	−0.0033	0.0033	−1.02	0.309
sup.imp	0.0072	0.872	−0.0025	0.0030	−0.84	0.398
sup.dom	−0.0586	0.190	−0.0053	0.0029	−1.83	0.068
autono3	−0.0771	0.085	−0.0866	0.0698	−1.24	0.215
equity.sh	−0.0199	0.656	0.0021	0.0026	0.80	0.422
d_know	−0.0945	0.034	−0.0723	0.0487	−1.49	0.137
d_market	0.0766	0.087	0.0842	0.0483	1.74	0.081
d_ client	0.0781	0.081	0.0905	0.0429	2.11	0.035
d_effic	0.0531	0.236	0.0549	0.0493	1.11	0.266
d_nat.res	0.1570	0.000	0.1475	0.0455	3.24	0.001
age	−0.0211	0.637	−0.0038	0.0121	−0.31	0.755
d_greenfield	0.1030	0.021	0.3319	0.1072	3.10	0.002
size	0.0140	0.755	0.0001	0.0001	0.73	0.465
d_da industry	0.1186	0.008	0.8139	0.3768	2.16	0.031
d_db industry	0.0694	0.121	0.4922	0.3691	1.33	0.182
d_dd industry	0.0198	0.658	0.1767	0.4993	0.35	0.723

(*Continued*)

Table A1   Continued

Variable	Partial correlation		Ordered probit analysis			
	Corr.	Sig.	Coef.	Robust Std. Err.	z	P>|z|
d_de industry	0.0506	0.258	0.4014	0.3876	1.04	0.300
d_dg industry	−0.0020	0.965	−0.0308	0.4120	−0.07	0.940
d_dh industry	0.0596	0.183	0.4091	0.4058	1.01	0.313
d_di industry	**0.0747**	**0.095**	**0.6352**	**0.3713**	**1.71**	**0.087**
d_dj industry	0.0410	0.360	0.2790	0.3688	0.76	0.449
d_dk industry	0.0547	0.222	0.2953	0.3643	0.81	0.418
d_dl industry	0.0651	0.146	0.4765	0.3547	1.34	0.179
d_dm industry	**0.1056**	**0.018**	**0.8525**	**0.3929**	**2.17**	**0.030**
d_dn industry	**0.0837**	**0.061**	**0.6501**	**0.3831**	**1.70**	**0.090**

Number of obs = 534

Number of obs = 522

Wald chi2(34) = 263.20   Prob > chi2 = 0.000

Log pseudolikelihood = −667.2520   Pseudo R2 = 0.1889

Table A2   IPR gaps and foreign affiliate technological activity: partial correlation results with *sh.R&D* and *sal.inno*

Variable	sh.R&D		Variable	sal.inno	
	Corr.	Sig.		Corr.	Sig.
IPR.gap	**−0.0877**	**0.073**	**IPR.gap**	**−0.1073**	**0.021**
tt.mnc	0.0044	0.929	**tt.mnc**	**0.0890**	**0.055**
own.tech	**0.2561**	**0.000**	**own.tech**	**0.1974**	**0.000**
HQ.tech	**0.0872**	**0.075**	HQ.tech	−0.0452	0.330
FA.tech	−0.0394	0.421	FA.tech	0.0340	0.464
for.SI	−0.0406	0.408	for.SI	−0.0397	0.392
dom.SI	0.0413	0.399	**dom.SI**	**0.1066**	**0.021**
sal.HQ	0.0054	0.912	sal.HQ	0.0671	0.148
sal.exp	−0.0573	0.242	sal.exp	−0.0252	0.587
sal.dom	−0.0733	0.134	sal.dom	0.0453	0.330

(Continued)

Table A2   Continued

	sh.R&D			sal.inno	
Variable	**Corr.**	**Sig.**	**Variable**	**Corr.**	**Sig.**
sup.HQ	**−0.0921**	**0.060**	sup.HQ	0.0097	0.834
sup.imp	−0.0049	0.921	sup.imp	−0.0148	0.750
sup.dom	−0.0301	0.538	sup.dom	−0.0510	0.272
autono3	**0.1147**	**0.019**	autono3	0.0038	0.935
equity.sh	0.0031	0.949	equity.sh	0.0182	0.695
d_know	0.0686	0.161	dom knowl strategy	0.0543	0.242
d_market	−0.0424	0.386	market acc strategy	0.0448	0.335
d_ client	0.0768	0.116	follow clien strategy	0.0012	0.979
d_effic	**−0.0825**	**0.092**	efficiency strategy	−0.0636	0.171
d_nat.res	−0.0137	0.779	nat ress strategy	−0.0303	0.514
age	0.0041	0.933	age	−0.0267	0.565
d_greenfield	−0.0149	0.761	d_greenfield	−0.0625	0.178
size	−0.0211	0.667	size	−0.0001	0.998
d_da industry	−0.0443	0.366	d_da industry	0.0415	0.371
d_db industry	**−0.1201**	**0.014**	d_db industry	0.0070	0.880
d_dd industry	**−0.0839**	**0.086**	d_dd industry	0.0197	0.672
d_de industry	−0.0694	0.156	d_de industry	0.0168	0.718
d_dg industry	−0.0341	0.486	d_dg industry	0.0586	0.207
d_dh industry	−0.0632	0.196	**d_dh industry**	**0.0880**	**0.058**
d_di industry	−0.0404	0.409	d_di industry	0.0002	0.996
d_dj industry	−0.0688	0.160	d_dj industry	0.0544	0.241
d_dk industry	−0.0081	0.868	d_dk industry	0.0733	0.114
d_dl industry	0.0014	0.977	**d_dl industry**	**0.0932**	**0.044**
d_dm industry	−0.0266	0.587	**d_dm industry**	**0.0962**	**0.038**
d_dn industry	0.0092	0.851	d_dn industry	0.0697	0.133
	Number of obs = 453			Number of obs = 500	

Table A3a IPR gaps and importance of sources of knowledge and technology for foreign affiliate technological activity: partial correlation and ordered probit results with *own.tech*

Variable	Partial correlation		Ordered probit analysis			
	Corr.	Sig.	Coef.	Robust Std. Err.	z	P>|z|
IPR.gap	−0.1233	0.006	−0.1302	0.0322	−4.05	0.000
tt.mnc	0.1218	0.006	0.1539	0.0558	2.76	0.006
HQ R&D	0.1419	0.001	0.1709	0.0559	3.06	0.002
HQ.tech	0.0611	0.172	0.0648	0.0523	1.24	0.215
for.SI	0.0730	0.103	0.2035	0.0712	2.86	0.004
dom.SI	0.1691	0.000	0.1734	0.0524	3.31	0.001
sal.HQ	0.0115	0.798	−0.0002	0.0033	−0.06	0.949
sal.exp	0.1247	0.005	0.0054	0.0033	1.67	0.094
sal.dom	0.0339	0.449	0.0017	0.0031	0.56	0.577
sup.HQ	−0.0055	0.902	−0.0001	0.0032	−0.02	0.982
sup.imp	−0.0009	0.984	0.0016	0.0030	0.53	0.598
sup.dom	0.0304	0.497	0.0031	0.0027	1.16	0.246
autono3	0.3032	0.000	0.4445	0.0681	6.52	0.000
equity.sh	−0.0031	0.945	−0.0015	0.0025	−0.61	0.540
d_know	0.1040	0.020	0.0850	0.0438	1.94	0.052
d_market	0.0077	0.863	−0.0065	0.0478	−0.14	0.891
d_ client	−0.1023	0.022	−0.0803	0.0425	−1.89	0.059
d_effic	0.0903	0.043	0.0811	0.0471	1.72	0.085
d_nat.res	−0.0567	0.205	−0.0782	0.0419	−1.87	0.062
age	0.0565	0.207	0.0081	0.0115	0.71	0.480
d_greenfield	−0.0658	0.141	−0.1680	0.1056	−1.59	0.112
size	−0.0510	0.255	−0.0001	0.0001	−0.79	0.429
d_da industry	−0.0594	0.185	−0.4248	0.3777	−1.12	0.261
d_db industry	−0.0730	0.103	−0.5887	0.3745	−1.57	0.116
d_dd industry	−0.0374	0.404	−0.4358	0.4405	−0.99	0.322
d_de industry	−0.0183	0.683	−0.2108	0.3950	−0.53	0.594
d_dg industry	−0.0538	0.230	−0.3993	0.3888	−1.03	0.304

(Continued)

Table A3a Continued

Variable	Partial correlation		Ordered probit analysis			
	Corr.	Sig.	Coef.	Robust Std. Err.	z	P>|z|
d_dh industry	−0.0214	0.632	−0.1528	0.3839	−0.40	0.691
d_di industry	−0.0676	0.131	−0.5336	0.3814	−1.40	0.162
d_dj industry	−0.0688	0.124	−0.4571	0.3722	−1.23	0.219
d_dk industry	−0.0125	0.780	−0.0336	0.3690	−0.09	0.928
d_dl industry	−0.0666	0.137	−0.5261	0.3639	−1.45	0.148
d_dm industry	**−0.0836**	**0.061**	**−0.6795**	**0.3842**	**−1.77**	**0.077**
d_dn industry	−0.0407	0.363	−0.3146	0.3790	−0.83	0.406

Number of obs = 534	Number of obs = 522	
	Wald chi2(34) = 229.56	Prob > chi2 = 0.000
	Log pseudolikelihood = −704.5254	Pseudo R2 = 0.1440

Table A3b IPR gaps and importance of sources of knowledge and technology for foreign affiliate technological activity: partial correlation and ordered probit results with *comp.HQ*

Variable	Partial correlation		Ordered probit analysis			
	Corr.	Sig.	Coef.	Robust Std. Err.	z	P>|z|
IPR.gap	**−0.1625**	**0.000**	**−0.1222**	**0.0324**	**−3.77**	**0.000**
tt.mnc	**−0.1556**	**0.000**	**−0.1364**	**0.0470**	**−2.90**	**0.004**
HQ.tech	**−0.1281**	**0.004**	**−0.1188**	**0.0539**	**−2.20**	**0.028**
for.SI	−0.0092	0.837	0.0319	0.0669	0.48	0.633
dom.SI	**0.1022**	**0.022**	**0.1116**	**0.0554**	**2.01**	**0.044**
sal.HQ	0.0174	0.698	0.0000	0.0028	0.01	0.990
sal.exp	**0.1414**	**0.001**	**0.0053**	**0.0028**	**1.86**	**0.063**

(Continued)

Table A3b   Continued

Variable	Partial correlation		Ordered probit analysis			
	Corr.	Sig.	Coef.	Robust Std. Err.	z	$P>|z|$
sal.dom	0.0350	0.434	0.0009	0.0025	0.37	0.712
sup.HQ	−0.0732	0.102	−0.0039	0.0029	−1.32	0.187
sup.imp	0.0254	0.570	0.0010	0.0029	0.35	0.723
sup.dom	0.0070	0.877	0.0009	0.0025	0.37	0.709
autono3	**0.3985**	**0.000**	**0.5459**	**0.0631**	**8.66**	**0.000**
equity.sh	−0.0533	0.233	−0.0043	0.0023	−1.91	0.056
d_know	0.0621	0.165	0.0447	0.0412	1.09	0.278
d_market	−0.0383	0.391	−0.0129	0.0443	−0.29	0.771
d_ client	−0.0638	0.153	−0.0582	0.0415	−1.40	0.161
d_effic	**0.0755**	**0.091**	0.0545	0.0442	1.23	0.218
d_nat.res	−0.0210	0.639	−0.0516	0.0404	−1.28	0.201
age	0.0170	0.705	0.0028	0.0104	0.27	0.787
d_greenfield	−0.0540	0.227	−0.1572	0.1005	−1.56	0.118
size	−0.0269	0.548	0.0000	0.0001	0.01	0.995
d_da industry	−0.0298	0.506	−0.1981	0.2864	−0.69	0.489
d_db industry	−0.0300	0.502	−0.2610	0.3009	−0.87	0.386
d_dd industry	−0.0101	0.822	−0.1391	0.3364	−0.41	0.679
d_de industry	−0.0066	0.882	−0.1041	0.3115	−0.33	0.738
d_dg industry	−0.0526	0.239	−0.4305	0.3365	−1.28	0.201
d_dh industry	−0.0139	0.755	−0.1130	0.2922	−0.39	0.699
d_di industry	−0.0161	0.718	−0.1290	0.2982	−0.43	0.665
d_dj industry	−0.0267	0.550	−0.2080	0.2857	−0.73	0.467
d_dk industry	0.0040	0.929	0.0871	0.2951	0.30	0.768
d_dl industry	−0.0483	0.280	−0.3635	0.2805	−1.30	0.195
d_dm industry	−0.0444	0.321	−0.3993	0.3153	−1.27	0.205
d_dn industry	−0.0079	0.860	−0.1510	0.2979	−0.51	0.612

Number of obs = 534

Number of obs = 522

Wald chi2(34) = 324.06   Prob > chi2 = 0.000

Log pseudolikelihood = −860.43414   Pseudo R2 = 0.1442

Table A3c IPR gaps and importance of sources of knowledge and technology for foreign affiliate technological activity: partial correlation and ordered probit results with *comp.FA*

Variable	Partial correlation		Ordered probit analysis			
	Corr.	Sig.	Coef.	Robust Std. Err.	z	P>|z|
IPR.gap	−0.1078	0.016	−0.0827	0.0308	−2.69	0.007
tt.mnc	0.0322	0.471	0.0100	0.0519	0.19	0.847
HQ R&D	−0.0513	0.252	−0.0573	0.0561	−1.02	0.308
for.SI	−0.0444	0.321	0.0086	0.0656	0.13	0.896
dom.SI	0.0669	0.134	0.0597	0.0509	1.17	0.241
sal.HQ	0.0131	0.770	0.0002	0.0036	0.05	0.963
sal.exp	0.0928	0.038	0.0043	0.0035	1.21	0.226
sal.dom	0.0256	0.567	0.0017	0.0034	0.49	0.627
sup.HQ	−0.0208	0.643	−0.0004	0.0036	−0.11	0.909
sup.imp	0.0125	0.780	0.0012	0.0034	0.36	0.718
sup.dom	0.0360	0.421	0.0021	0.0032	0.65	0.517
autono3	0.2563	0.000	0.3324	0.0609	5.46	0.000
equity.sh	0.0057	0.898	−0.0017	0.0024	−0.73	0.468
d_know	0.0498	0.266	0.0459	0.0409	1.12	0.262
d_market	−0.0413	0.356	−0.0465	0.0436	−1.07	0.286
d_ client	−0.0820	0.066	−0.0640	0.0396	−1.61	0.106
d_effic	0.0915	0.040	0.0951	0.0450	2.11	0.034
d_nat.res	−0.0673	0.132	−0.0634	0.0406	−1.56	0.118
age	0.0842	0.059	0.0157	0.0103	1.53	0.126
d_greenfield	−0.0151	0.737	−0.0150	0.0997	−0.15	0.880
size	−0.0507	0.256	−0.0001	0.0001	−0.73	0.465
d_da industry	−0.0094	0.834	−0.0566	0.3393	−0.17	0.868
d_db industry	−0.0281	0.530	−0.2969	0.3133	−0.95	0.343
d_dd industry	−0.0063	0.888	−0.0862	0.4017	−0.21	0.830
d_de industry	0.0007	0.987	0.0222	0.3502	0.06	0.950
d_dg industry	−0.0282	0.529	−0.3010	0.3650	−0.82	0.410
d_dh industry	0.0151	0.737	0.1575	0.3522	0.45	0.655
d_di industry	−0.0648	0.147	−0.4464	0.3293	−1.36	0.175

(Continued)

Table A3c   Continued

Variable	Partial correlation		Ordered probit analysis			
	Corr.	Sig.	Coef.	Robust Std. Err.	z	P>|z|
d_dj industry	−0.0418	0.350	−0.3491	0.3281	−1.06	0.287
d_dk industry	0.0012	0.978	−0.0095	0.3345	−0.03	0.977
d_dl industry	−0.0200	0.655	−0.2131	0.3162	−0.67	0.500
d_dm industry	−0.0550	0.219	−0.5368	0.3482	−1.54	0.123
d_dn industry	−0.0213	0.635	−0.1869	0.3410	−0.55	0.584

Number of obs = 534   Number of obs = 522

Wald chi2(34) = 128.66   Prob > chi2 = 0.000

Log pseudolikelihood = −876.24311   Pseudo R2 = 0.0674

Table A4a IPR gaps and technological embeddedness in the local host economy:partial correlation and log-OLS results with *comp.sal*

Variable	Partial correlation		Ordered probit analysis			
	Corr.	Sig.	Coef.	Robust Std. Err.	t	P>|t|
IPR.gap	**0.0883**	**0.048**	**1.4851**	**0.697**	**2.13**	**0.034**
tt.mnc	0.0640	0.152	0.0279	0.023	1.21	0.225
own.tech	−0.0679	0.128	−0.0365	0.021	−1.73	0.085
HQ tech	0.0320	0.474	0.0144	0.024	0.59	0.553
HQ.tech	0.0291	0.515	0.0085	0.021	0.40	0.688
for.SI	−0.0646	0.148	−0.0093	0.025	−0.37	0.712
dom.SI	0.0660	0.139	0.0015	0.022	0.07	0.945
sup.HQ	**0.2188**	**0.000**	**0.2395**	**0.084**	**2.85**	**0.005**
sup.imp	**−0.1221**	**0.006**	−0.1178	0.078	−1.51	0.131
sup.dom	−0.0695	0.120	−0.1055	0.079	−1.33	0.183
autono3	−0.0776	0.082	−0.0423	0.028	−1.52	0.128

(Continued)

equity.sh	0.0901	0.043	0.0594	0.027	2.18	0.029
d_know	0.1686	0.000	−0.1159	0.022	−5.29	0.000
d_market	−0.2149	0.000	−0.0416	0.019	−2.17	0.031
d_ client	−0.0995	0.026	0.0877	0.022	4.04	0.000
d_effic	0.1806	0.000	−0.0008	0.019	−0.04	0.966
d_nat.res	−0.0155	0.729	0.0719	0.021	3.50	0.001
age	−0.0584	0.191	−0.0224	0.019	−1.20	0.230
d_greenfield	−0.0778	0.081	−0.0356	0.022	−1.63	0.103
size	0.0656	0.142	0.0119	0.008	1.48	0.139
d_da industry	−0.1120	0.012	−0.1909	0.057	−3.32	0.001
d_db industry	−0.0439	0.325	−0.0709	0.058	−1.22	0.225
d_dd industry	0.0681	0.127	0.0834	0.070	1.19	0.234
d_de industry	−0.0404	0.366	−0.1437	0.062	−2.32	0.021
d_dg industry	−0.0612	0.171	−0.1371	0.064	−2.15	0.032
d_dh industry	−0.0022	0.961	−0.0339	0.068	−0.50	0.616
d_di industry	−0.0681	0.127	−0.1470	0.058	−2.52	0.012
d_dj industry	−0.0268	0.549	−0.0480	0.057	−0.84	0.401
d_dk industry	−0.0601	0.178	−0.1097	0.058	−1.88	0.060
d_dl industry	−0.0320	0.474	−0.0572	0.056	−1.02	0.308
d_dm industry	−0.0867	0.052	−0.1644	0.065	−2.53	0.012
d_dn industry	−0.0630	0.158	−0.1116	0.065	−1.70	0.089
constant			−2.1699	3.390	−0.64	0.522

Number of obs = 534	Number of obs = 522
	F(32, 489) = 17.44 Prob > F = 0.0000
	Root MSE = 0.22149 R-squared = 0.3809

Table A4b IPR gaps and technological embeddedness in the local host economy: partial correlation and log-OLS results with *comp.sup*

Variable	Partial correlation		Ordered probit analysis			
	Corr.	Sig.	Coef.	Robust Std. Err.	t	P>|t|
IPR.gap	0.0600	0.179	0.2469	0.6975	0.35	0.724
tt.mnc	**0.0745**	**0.095**	**0.0361**	**0.0218**	**1.66**	**0.098**
own.tech	−0.0273	0.542	−0.0377	0.0230	−1.64	0.101
HQ tech	0.0676	0.130	0.0269	0.0242	1.11	0.267
HQ.tech	0.0505	0.258	0.0038	0.0206	0.18	0.856
for.SI	0.0660	0.140	**0.0440**	**0.0252**	**1.74**	**0.082**
dom.SI	**−0.1119**	**0.012**	**−0.0517**	**0.0222**	**−2.33**	**0.020**
sal.HQ	**0.1329**	**0.003**	−0.0025	0.0860	−0.03	0.977
sal.exp	**−0.2237**	**0.000**	**−0.3913**	**0.0862**	**−4.54**	**0.000**
sal.dom	−0.0635	0.155	**−0.2173**	**0.0856**	**−2.54**	**0.011**
autono3	**−0.1158**	**0.009**	−0.0569	0.0324	−1.76	0.079
equity.sh	0.0394	0.378	0.0159	0.0268	0.59	0.552
d_know	−0.0594	0.183	−0.0029	0.0229	−0.13	0.900
d_market	0.0046	0.918	**0.0397**	**0.0175**	**2.27**	**0.023**
d_ client	**0.0971**	**0.030**	0.0141	0.0213	0.66	0.507
d_effic	0.0296	0.508	**−0.0376**	**0.0181**	**−2.08**	**0.038**
d_nat.res	**−0.0986**	**0.027**	−0.0232	0.0193	−1.20	0.230
age	−0.0039	0.930	−0.0076	0.0179	−0.42	0.673
d_greenfield	−0.0480	0.282	−0.0004	0.0219	−0.02	0.986
size	−0.0365	0.414	−0.0095	0.0082	−1.15	0.249
d_da industry	**−0.0848**	**0.057**	**−0.1419**	**0.0652**	**−2.18**	**0.030**
d_db industry	−0.0729	0.102	−0.1049	0.0644	−1.63	0.104
d_dd industry	**−0.0867**	**0.052**	**−0.2016**	**0.0715**	**−2.82**	**0.005**
d_de industry	−0.0524	0.241	−0.1003	0.0789	−1.27	0.204
d_dg industry	−0.0481	0.282	−0.0919	0.0688	−1.34	0.182
d_dh industry	−0.0435	0.330	−0.1274	0.0699	−1.82	0.069
d_di industry	**−0.0912**	**0.041**	**−0.1589**	**0.0656**	**−2.42**	**0.016**
d_dj industry	**−0.0743**	**0.096**	**−0.1275**	**0.0646**	**−1.97**	**0.049**

(Continued)

Table A4b   Continued

Variable	Partial correlation		Ordered probit analysis			
	Corr.	Sig.	Coef.	Robust Std. Err.	t	P>|t|
d_dk industry	−0.0569	0.203	−0.1025	0.0632	−1.62	0.105
d_dl industry	−0.0436	0.330	−0.0919	0.0602	−1.53	0.127
d_dm industry	−0.0517	0.247	**−0.1105**	**0.0643**	**−1.72**	**0.086**
d_dn industry	−0.0464	0.299	−0.1109	0.0702	−1.58	0.115
constant			6.7951	3.3869	2.01	0.045

Number of obs = 534

Number of obs = 522

$F(32, 489) = 7.37$ Prob > F = 0.0000

Root MSE = 0.21323   R–squared = 0.2876

Table A4c   IPR gaps and technological embeddedness in the local host economy: partial correlation and ordered probit results with *comp.SI*

Variable	Partial correlation		Ordered probit analysis			
	Corr.	Sig.	Coef.	Robust Std. Err.	z	P>|z|
IPR.gap	**−0.0847**	**0.061**	**−0.0587**	**0.0345**	**−1.70**	**0.089**
tt.mnc	**0.1302**	**0.004**	**0.1301**	**0.0476**	**2.74**	**0.006**
HQ.tech	0.0449	0.321	0.0509	0.0438	1.16	0.244
for.SI	−0.0596	0.188	−0.0517	0.0491	−1.05	0.292
dom.SI	−0.0142	0.754	−0.0225	0.0481	−0.47	0.639
sal.HQ	0.0310	0.494	0.0004	0.0004	1.03	0.305
sal.exp	−0.0049	0.914	−0.0007	0.0028	−0.25	0.800
sal.dom	0.0430	0.343	0.0033	0.0025	1.32	0.187
sup.HQ	−0.0465	0.305	−0.0042	0.0030	−1.41	0.157
sup.imp	−0.0152	0.737	−0.0009	0.0030	−0.31	0.753

(Continued)

Table A4c Continued

Variable	Partial correlation		Ordered probit analysis			
	Corr.	Sig.	Coef.	Robust Std. Err.	z	P>|z|
sup.dom	0.0423	0.350	0.0021	0.0027	0.77	0.442
autono3	0.0523	0.248	0.0705	0.0647	1.09	0.276
equity.sh	0.0327	0.470	0.0005	0.0024	0.19	0.847
d_know	−0.0048	0.916	0.0086	0.0423	0.20	0.839
d_market	0.0171	0.705	0.0038	0.0431	0.09	0.930
d_ client	−0.0172	0.705	−0.0084	0.0408	−0.21	0.836
d_effic	−0.0218	0.631	−0.0028	0.0444	−0.06	0.949
d_nat.res	0.0190	0.674	−0.0059	0.0412	−0.14	0.887
age	0.0147	0.746	0.0036	0.0111	0.33	0.743
d_greenfield	−0.0685	0.130	−0.1727	0.1054	−1.64	0.101
size	0.0195	0.667	0.0000	0.0001	−0.43	0.669
d_da industry	−0.0119	0.793	−0.1241	0.3445	−0.36	0.719
d_db industry	0.0234	0.606	0.1187	0.3406	0.35	0.727
d_dd industry	−0.0507	0.263	−0.4546	0.4136	−1.10	0.272
d_de industry	−0.0258	0.570	−0.3039	0.3924	−0.77	0.439
d_dg industry	0.0449	0.321	0.3740	0.3588	1.04	0.297
d_dh industry	0.0248	0.584	0.1671	0.3698	0.45	0.651
d_di industry	0.0452	0.318	0.3208	0.3525	0.91	0.363
d_dj industry	0.0473	0.296	0.3425	0.3383	1.01	0.311
d_dk industry	0.0629	0.165	0.4245	0.3393	1.25	0.211
d_dl industry	0.0402	0.374	0.3233	0.3329	0.97	0.332
d_dm industry	0.0418	0.356	0.3433	0.3586	0.96	0.338
d_dn industry	−0.0052	0.909	−0.0708	0.3617	−0.20	0.845

Number of obs = 522

Number of obs = 522

LR chi2(34) = 66.27   Prob > chi2 = 0.0005

Log pseudolikelihood = −633.58669   Pseudo R2 = 0.0497

8
Conclusions on the Technological Role of FDI into CEE

Before economies that are catching up in terms of technological development can feed the process of technical advance from their own indigenous resources, they will typically depend on the supply of technology from abroad, that is technological diffusion and imitation. FDI is the most important driver of supply of external technology. This is generally assumed for the newly industrialised countries in Asia and other emerging markets, and has also proven to be the case amongst some of the West European countries in their own processes of catching up (see for example Jungmittag, 2005). There is, however, a controversy about whether FDI has in fact had a benign effect for economic development or not. Even if the literature does on average agree that FDI has positively contributed to technological advance and economic growth (for comprehensive overviews see for example Moran et al., 2005; Rugman and Doh, 2008), many important qualifications can be found in the literature that are convincing and appear to be empirically robust. The final word on the developmental role of TNCs is hence not out yet, and may never be due to issues like heterogeneity and context-specificity.

The countries in Central East Europe also constitute emerging markets, hence the widespread hope that FDI could potentially help to improve the conditions in CEECs for catching up in terms of technology. At the most general level, this obviously depends on whether FDI inflows are substantial and on whether the effects are benign or malign. Whilst the amount of inward FDI in CEE can be measured and tested in terms of significance for economic development (growth contribution of investment made and employment created by FDI, etc.), the technology issue is empirically not testable to an unambiguous result and hence not predictable. Nor is there a sufficient theory, nor a formal model that allows the factors of technology and knowledge transfer and spillovers to be determined and empirically tested. This work has therefore assessed the conditions that have to be fulfilled so that inward FDI can have a positive and important role in the technological process of catching up in CEE. These conditions are identified in the relevant literature, which includes both theoretical deductions of (nomological) rules and/or assumptions (mainly in the resource-based view, but also the systemic innovation literature to some extent) and plausibility assumptions based on the rationality of managers (mainly in the strategic management literature). This literature is organised in different topical foci, each assessed in the five research chapters of this work: the strategic motives of investment and their match with locational conditions; internal technology and knowledge transfer within the network of the foreign investment; external technology and knowledge sourcing of foreign affiliates from their host innovation system; technology transfer and spillovers into the host economy innovation system; and the role of the technology-related institution of IPR regime for technology transfer and spillovers.

The relevance of determinants was then tested empirically by use of the unique IWH FDI Micro Database. The tests either found statistically significant support, or rejected the significance of determinants hypothesised to drive a positive developmental role of FDI, or produced insignificant results. Each of the tests was conducted in its own setting of assumed relevant determinants (the topical foci), which still led to a number of overlaps. Methodologically, there is no proof as to the completeness of conditions and determinants. Moreover, empirically significant results can only be interpreted to find support for a hypothesis, without being able to offer proof. And whilst statistically insignificant (or significant but counterintuitive) results are able to refute the respective hypothesis, this is so only in their own setting: they are not able to grant for all times and cases when the hypothesis is wrong.

8.1   Conditions for inward FDI in CEE to have a developmental impact via technology

If FDI were able to serve as panacea for transition and catching-up development in CEECs (the overarching research question), then a set of conditions has to be fulfilled. This work assumes a set of four groups of conditions to be particularly important. These were derived from theoretical and empirical literature and have been tested in this work at the firm level: the extent or quality of a positive technological role of inward FDI in CEECs is deemed to depend on (i) the match between strategic motives of the investor and locational conditions; (ii) the group of conditions related to internal network technology transfer and spillovers; (iii) the group of conditions related to external technology transfer and spillovers; and (iv) the quality of formal institutions and their enactment as its informal pendent. Of course, other conditions may also play a role; this list is open – yet the four groups of determinants are treated as the important conditions in the relevant literature.

8.1.1 The match between strategic motives and locational conditions

In terms of strategic motives in CEECs, cost advantages (mainly due to lower unit labour costs) as well as market access (including the markets in the East: Russia and the so-called Community of Independent States; and the markets in the West: the EU common market) are the dominant reasons why foreigners invest in this region. Access to localised technology is of less importance and has not become more important since the outset of systemic transition. More important than the strategic motive itself, however, will be the match between the individual motive(s) of a foreign investor and the investor's perception of locational advantages and possibly potentials for development thereof: only where foreign investment projects find the kind of locational conditions and potentials that correspond to their expectations will FDI be able to be successful and thereby generate further investment. This new investment will to some extent again target the identified locational advantages, thereby reinforcing the former. This cumulative processes works at three different levels: at the regional level, where the region benefits from re-investments and gives rise to agglomeration advantages; at the sectoral-industry level, where the industry of the successful FDI project benefits from recurrent investment and strengthening locational conditions (which eventually may give rise to the development of a cluster); and at the macro-level, where the reallocation of resources resulting from re-investments aligns the countries' patterns of specialisation with the existing and emerging comparative advantages of the host economy. This conceptualisation is a reflection of a potentially active nature of FDI and their affiliates with respect to locational advantages in Cantwell's view of the world.

In a comparison between countries, the empirical analysis suggests that positive conditions for a technological role of inward FDI exist where the motive of reaping cost advantages is matched with an ample supply of labour at all levels of education and skills at low costs, as well as the availability of state support. In fact, such positive matches can be assumed for Poland with respect to the supply with employees at medium levels of qualification and in Hungary with respect to the supply of young university graduates. These countries can hence expect their human capital endowment strength to be well appreciated by foreign investors, and this may subsequently develop into a comparative advantage for the country. With regard to state support, a positive match is found in Hungary. The country may hence mainly build on the contemporary strength of providing state support for foreign investors (not necessarily the sheer size of it, but the match with expectations and supply), or else attempt to improve other determinants.

Finally, a positive match between local-technology seekers and potentials for local technological cooperation is found in the case of the Czech Republic, where collaboration with scientific institutions is sought, and in the case of Slovakia for collaboration with other domestic firms. Those countries can hence expect their comparative advantages of offering an able and attractive innovation system for FDI affiliates to solidify and possibly to develop into a comparative advantages as a host location for inward FDI.

8.1.2 Group of conditions related to internal technology transfer and spillovers

Internal technology transfer and spillovers between the foreign affiliate and the foreign investor's network (headquarters and other affiliates) constitute the second group of conditions for inward FDI to have a positive developmental role in terms of technology. Here, it is not, as often viewed, the knowledge and technology that flows only from parent to foreign affiliate; where the foreign investor (with its own network of foreign investments in other countries) is able to benefit from technology and knowledge accumulated in its foreign affiliate, technology transfer and spillovers hence run both ways, and are reciprocal. In such a case, the rate and value of technological advance within the whole network will be higher. The potential benefits for the host economy are hence also larger.

There are several sub-conditions attached to this: for internal transfer and spillovers to be particularly fruitful, the foreign affiliate will have a lower level of productivity *vis-à-vis* its foreign investor (Gerschenkron's 'advantages of backwardness'), but the gap must not be too large (insufficient absorptive capacities). In the case of all CEECs, there are significant gaps in technology, but they are clearly well above the gaps to be observed in other emerging markets or even developing countries (that also receive inward FDI). The gaps can be assumed to be sufficiently low to allow firms in CEECs to be able to absorb the foreign technology and to adapt it to function well in their own particular environment. The endowment with a well-trained workforce with an industrial history and its own R&D in CEECs (even if still low in aggregate data) is an indication of this. Following Polanyi's idea (that the owner of knowledge and technology often knows more than he is able to tell), a related condition is that all partners in the foreign investor's network additionally have to be aware of the tacit part of the knowledge and be able to convey this: all organisations in the network have to be able to learn, be they at the sending or receiving end. Directly emanating from this is the condition that was termed above as the level of autonomy of the foreign affiliate *vis-à-vis* the foreign investor. Again, no linear relationship can be found – the extent of developmental impact of inward FDI via internal technology transfer depends on the relationship between autonomy and absorptive capacities. It is highest where foreign affiliates are both highly autonomous and have a well-developed capability to make good use of the foreign knowledge and technology. According to the taxonomy suggested in Chapter 4 of this work, it is first and foremost East Germany that fulfils the condition of such a two-way dynamic internal technology and knowledge transfer, followed by Romania and, with a larger gap, the other countries, with Poland at the bottom of the list.

Further, the kind of knowledge and technology that is transferred between the network partners plays an important role. Here, the transfer of R&D involves the participation of CEEC's foreign affiliates in the general R&D strategy of the foreign investor, and this will have more technological impact than the diffusion of established knowledge and technology, however valuable that may be. Via R&D, the knowledge and technology that is exchanged is evolving, and hence represents the state of the art. Moreover, technology generated with the active participation of foreign affiliates in CEECs will tend to be more easily applied in the CEEC environment itself than will technology developed somewhere else without reference to CEECs. Due to its industrial history both before and during the socialist era, the industries in CEECs are able to contribute to the world's technological standards; but alas the financial strengths of its firms and the comparatively less capitalised financial and capital markets make the embedding of their firms' R&D into foreign ownership networks an important condition. The IWH FDI Micro Databases do identify that such R&D participation exists for foreign affiliates in CEECs, even if on a comparatively low scale.

Finally, the state of development of formal institutions (like the legal framework) are an important condition for sensitive knowledge and technology to be able to flow between foreign investor network and foreign affiliate without the threat of loss of ownership advantage. Despite the fact that CEECs are either already included in the *acquis communautaire* of the EU or voluntarily take over those regulations, their intensity of enforcement (as an informal institution within the norms and values of the society) is still weaker than for most foreign investors' home countries.

8.1.3   The group of conditions related to external technology transfer and spillovers

The group of conditions related to external technology transfer and spillovers is multi-faceted: it depends on the conditions of technological activity of foreign affiliates in the first place. And they, furthermore, are reminiscent of foreign investment strategies, of absorptive capacities and technological capabilities of foreign affiliates, of the conditions for collaboration between foreign affiliates and actors in their host innovation system, and finally of the technological capabilities of the host innovation system actors themselves[1]. Complicating matters, those conditions may differ between the two channels for a positive technological role (i) of local knowledge and technology sourcing by foreign affiliates (where the positive technological role emanates from the technical advance of the foreign affiliate itself and the improvement of the location for further investment and local economic activity searching for a vital and dynamic knowledge base). And the positive technological role that roots in (ii) local firms benefiting from technology transfer and spillovers originating from foreign affiliates in their location (which is the obvious and straightforward channel for such a positive technological role).

The condition of R&D and innovation-related technological activity of foreign affiliates in CEECs is fulfilled comparatively well considering the low levels of economic development in these countries. This is not least a result of their industrial history – an advantage that most newly industrialised economies do not have. In comparison with the other economies further east that formally were part of the Soviet Union, CEECs are particularly well placed to experience their foreign affiliates engaging in technological activity, because of the high level of FDI penetration and the intense integration into the Western European knowledge and technology industries.

With regard to strategies of FDI projects, the analysis established that it is not only projects that explicitly search for new knowledge and technology beyond the existing knowledge base of the foreign investor's network (the augmenting strategy) that fulfil the condition for a positive technological role of inward FDI. Foreign investments that focus on exploiting the home-base knowledge stock are equally subject to intense collaboration with the host innovation system to assist in the foreign affiliates' own technological activity. If it is plausibly assumed that such technological collaboration is a driver of a positive technological role of inward FDI, then the distinction between those two strategies made in the literature is not relevant and FDI projects of either strategic character positively fulfil the condition of a positive technological role. Further, the analysis identifies the embeddedness of foreign affiliates into the local host economy in terms of local procurement and local sales as a condition for a positive technological role of inward FDI. This is because the intensity of local trade increases the propensity of technological collaboration with the respective local trade partner. This may be with a view on customisation of product and process technologies to the needs, preferences, possibilities and particularities of the local host market. The result that upstream and downstream trade relations drive local technological cooperation suggests a positive evaluation of the capabilities of the local economy in the case of CEECs. A capable host innovation system does in fact serve as a condition for a positive technological role of inward FDI by assisting foreign affiliates in their own technological efforts. This, moreover, suggests that the quality of human capital forms another related condition for a positive technological role of inward FDI. And here, the transition economies in CEE have a particularly important task ahead of them: first, formal qualifications have often borne little relevance to the actual needs in the current occupations of workers, whilst working experience was found unable to fully bridge the gap. Second, a proportion of personnel crucial to the raising of the value of human capital in CEECs often left domestic firms for better paid jobs in foreign affiliates or abroad (the brain drain). Finally, the importance attached to a capable host innovation system directs attention to the state of transformation and ability of local science institutions. Here again, CEECs have some way to go to make those actors in their innovation systems truly attractive for technology-intensive, or technology-seeking, inward FDI.

The ambiguous results in the relevant literature with respect to the corporate governance issue turns out to be straightforward for the case of CEECs: the more autonomous are foreign affiliates in deciding on their own business functions, the higher the propensity to source technology and knowledge from local sources. Autonomous firms are also more important sources of knowledge and technology for local suppliers. Hence, autonomy can be assumed to be a relevant condition for a positive technological role of inward FDI. With respect to strategic investment motives, the analysis suggests that the motive to tap localised knowledge and technology is not the only significant driver of local technological embeddedness. Rather, FDI projects following efficiency-related motives, even if targeting lower unit labour costs, are able to fulfil the condition that inward FDI can have a positive technological impact on its host economy. This is not at all an obvious finding, and provides an important piece of information for policies trying to attract inward FDI. Finally, the empirical results show that FDI into existing firms (for example via privatisation or merger & acquisition), that is, non-greenfield investments, are an important condition for a positive technological role from an embeddedness perspective[2], while the time of engagement in the host economy (liability of newness) and the size of the FDI project (trustworthiness) appear to be rather less relevant.

The analysis likewise does not find overwhelming empirical support for the pipeline concept of technology transfer and spillovers: in the selection of CEECs included in this analysis, internal technology transfer from the foreign investor network to the foreign affiliate is only important for knowledge and technology to spill over to local suppliers and not to customers downstream or even to local scientific institutions. Hence, internal technology transfer is not always an empirically significant pre-condition for technology spillovers in CEECs.

All the more, however, does the technological ability of the host innovation system play a role for spillovers in CEE: only where the local host innovation system is technologically active and able will spillovers be potentially sizeable. This suggests that a reciprocal exchange of knowledge and technology between the host innovation system and the foreign affiliate is a particularly important condition for a positive technological role of inward FDI.

The conditions derived from the insights generated in Chapter 5 emanate from criteria that allow foreign affiliates to excel in own technological activity (namely by local knowledge and technology sourcing). The conditions identified in Chapter 6, however, go one step further and may be viewed as necessary factors for the local host economy to benefit in terms of technology transfer and spillovers.

Because the above conditions were assessed by use of all sample countries in CEE in one panel, the analysis is unable to distinguish the relative relevance between the individual countries. But those conditions do differ between the countries: internal technology transfer turned out to be an important condition for external spillovers (the pipeline concept) in the group of Polish FDI projects – but not, however, in East Germany. This corresponds to the respective sizes of the two markets and the observation that Polish FDI is more focused on the domestic market whereas East German FDI more on exporting (including to West Germany). Consequently, it is also the Polish foreign affiliates that appear to produce more spillovers to local customers than any other country assessed. Own technological activity not only differs in general, but also with respect to the kinds of technological activity: Croatian and East German foreign affiliates better fulfil the condition of a positive technological role of FDI by spending more resources for R&D than FDI projects in Poland, Slovenia, and Romania. On the other hand, East German, Slovenian and Polish affiliates are more successful in marketing their innovations than Romanian and Croatian ones. Both criteria for this condition may be equally important: foreign affiliates' own R&D can spur the positive technological role by way of external transfer and spillover of newly generated tacit and explicit knowledge and technology. Innovative success by foreign affiliates can give rise to a positive technological role of inward FDI by way of diffusion of knowledge and technology embodied in the new products.

Whilst strategic investment motives are somewhat more homogeneous across the countries assessed here, Polish FDI projects are consistently less targeted at tapping the local host knowledge and technology base, whereas the projects in the Czech Republic, Slovakia, Croatia, East Germany, and Romania are more so. With regard to the second efficiency or cost-related motive likely to give rise to a positive technological role of inward FDI, Croatia and Romania are in a more comfortable position than the other countries.

Embeddedness of FDI projects with their host economies is also quite different between the countries assessed: in terms of trade relationships, Polish and East German foreign affiliates entertain particularly intense trade relations with local suppliers and local customers alike. Technological embeddedness with the host innovation system is particularly intense amongst Polish foreign affiliates (mostly with customers) and for both directions of knowledge flows (Chapters 5 and 6), whilst it is less intense for East German, Slovenian, and Croatian foreign affiliates with respect to a direction of knowledge flows from foreign affiliates to local firms, that is the typical spillover case (Chapter 6). East German and Romanian foreign affiliates are more intensively technologically embedded for a direction of knowledge flows from local firms to foreign affiliates (Chapter 5) than serving themselves as sources of knowledge and technology for local firms (Chapter 6). Both criteria for this condition of a positive technological role may again be equally important: whereas the Chapter 5 kind of criterion will attract more knowledge-seeking FDI projects and will allow any other kind of FDI project to reap advantages of technology sourcing, the spillover-related criterion of Chapter 6 is the channel traditionally thought of in terms of a positive developmental role of inward FDI via technology.

Next to the embeddedness issue, it is also the ability of the host innovation system to engage in reciprocal exchange of knowledge and technology that is found to be an important condition for a positive technological role of inward FDI. In a comparison between the host countries assessed in this analysis, it is the host innovation systems of primarily Poland, East Germany, and possibly also Croatia that are particularly well adapted to such a two-way exchange. Romanian and Slovenian FDI projects appear to be less able to fulfil this important condition for inward FDI to generate positive technological effects in the host economy.

8.1.4 The quality of the IPR regime as a formal institution

The final set of conditions is institutional and pertains to the quality of protection of intellectual property rights (IPR), which in conceptual terms form part of Dunning's ownership advantages of the FDI project. The analysis conducted here is able to disentangle the relationship between the strength of IPR protection and the technological role of inward FDI: where the gap in IPR protection is particularly large between the FDI project's home and host countries, foreign affiliates were both less technologically active as well as technologically less embedded into their respective host innovation systems. To put it more positively: a reliably strong protection of intellectual property rights, both in terms of formal institution (the law in the books) and in terms of informal norms and values (the enforcement possibilities and costs involved) are found to be important conditions for inward FDI to generate positive technological effects in the host economy.

In a comparison between countries, empirical analysis by use of the IWH FDI Micro Database of 2006–2007 and the WEF IPR data suggests that positive conditions for a technological role of inward FDI exist currently in those host economies that have low IPR gaps in their endowment with FDI projects (see Table 8.1). This ranks East Germany first, where IPR levels are on average higher than in the foreign

Table 8.1 WEF IPR levels of host countries and average sizes of IPR gaps between home and host countries, by groups of foreign affiliates according to their host countries

	WEF IPR level	IPR gap	sd	n
Croatia	3.7	+2.14	0.98	135
East Germany	6.6	−0.90	0.74	294
Poland	3.6	+2.48	0.63	102
Romania	3.1	+1.89	1.19	217
Slovenia	4.5	+1.46	0.88	40

Note: sd denotes standard deviation, n is the number of foreign affiliates.

Source: IWH FDI Micro Database 2006–2007; WEF (2007), Section IX, Table 3.3.

investors' home countries (a negative average gap for its FDI projects). Germany is followed by Slovenia, Romania, Croatia and finally Poland. This ranking does not necessarily have to correspond one-to-one with the country-specific WEF indicators for the quality of the IPR regime, because the former is determined by the country's own endowment with FDI projects from a specific set of home countries.[3]

In sum, the analysis shows that the host economies in CEE fulfil the sets of conditions established here to very different extents. Furthermore, and reminiscent of the heterogeneity issue, each host country fulfils different kinds or groups of conditions particularly well. Whilst this is not a result of a mechanism that would be related to specialisation on host country-specific advantages, the results may still be used by policymakers of these countries in an attempt to focus on the promotion of strengths in their own pattern of specialisation on fulfilment of conditions. Because the work here remains unable to determine the respective levels of importance of each of the conditions for a positive technological role identified (due to the lack of reliable proxies for the extent of actual positive effects), the analysis is also unable to predict from the shape of conditions which host economy in CEE is in sum best set to technologically benefit from inward FDI.

The conditions for inward FDI to have a developmental impact via technology are mainly to be found in the micro-sphere. The conditions identified in this work are necessarily at the firm level, because of the heterogeneity amongst foreign investors in terms of strategic motives, modes of entry, and the strategic organisational management of foreign affiliates in the host economy etc. Complicating matters, strategic motives are typically overlapping, and organisational management of foreign affiliates is an issue for both the investing firm and the foreign affiliate (see the autonomy issue, the developmental-subsidiary issue). This results in a set of conditions that do not lend themselves to developing a general theory, and necessitates the description of conditions in a methodological framework of a taxonomy of important criteria. The results lend themselves all the more to be used by policymakers with a view to improving the conditions that this analysis found to be important for a positive technological role of inward FDI.

8.2   The research programme – some vital ingredients into a future research agenda

The research presented here offers some important insights into the conditions of inward FDI into CEECs to have a positive technological effect on the host economies. The analysis, however, remains constrained in several respects, suggesting a variety of paths for a future research agenda on the overarching research question. Three main avenues are briefly discussed in the following and are concerned with the further development of the micro database and additional research foci of importance for the technological role of FDI, as well as extensions to the use of theories or concepts.

8.2.1   Extensions to the micro database

The IWH FDI Micro Database is unique for the case of FDI into CEECs. It offers firm-specific information that reflects the insights to be drawn from conceptual and theoretical frameworks for the explanation of technology transfer and spillovers. It covers a variety of heterogeneous countries with different levels of economic development, different sizes, different patterns of sectoral specialisation and different economic policies, yet they they make an insightful group due to their common industrial history within the socialist era. And the database can claim some representativeness for the industries of the countries included. The database was used in many empirical studies to test propositions derived from theory and provided important insights into knowledge and technology transfer and spillovers in this particular region.

Yet, some important issues that would warrant further analysis remain beyond the abilities of the database. Some of those are easily added to the database in principle, while for others, the level of the foreign affiliate as the subject of analysis restrains the epistemological capabilities of the database.[4]

The central character of the database of focusing exclusively on the foreign affiliate precludes the numerical calculation of technological effects of foreign affiliates on the host economy. If the effect of inward FDI on host economy firms could be measured at the firm level, then analysis could calculate the economic (not statistical) significance of each of the conditions individually and in their interaction, that is, to give a specific value to the conditions and sets of conditions tested so far. This would not only shed more light on the mechanisms at work in knowledge and technology transfer and spillover (which would be important information for local and foreign-owned businesses alike), but also, and not least, be of utmost importance for economic policy in its attempt to increase the technological benefit for its economy from hosting inward FDI. Such an extension of the database is possible in two ways: either by broadening of the fieldwork to local firms or by matching the IWH FDI Micro Database with existing firm-level data sources (for example provided by official statistics or by private rating agencies, such as the Amadeus database) on the foreign affiliates interrogated and on local firms. Obvious candidates for such an extension could include data on value added per employee, research and development, as well as profitability, a small number of additional data with potentially high additional explanatory power (see Fritsch et al., 2004). It would also allow the broadening of the scope of analysis to the effects of FDI on R&D and innovative activity of local firms, an issue that can be expected to become ever more important empirically as the host countries develop their own technological capacities over time. Any matching of fieldwork-generated firm-level data with data published in official statistics is, however, riddled with the legal problem of confidentiality (for the many technical and legal obstacles, see for example Wagner, 2010).

A second obvious and pressing extension of the database would be to extend the focus to the locational dimension. Complementary data on the immediate location of the foreign affiliates would allow the analysis to not only test the spatial proximity assumption in the analysis of embeddedness with actors in the host innovation system (mainly Chapters 5 and 6), but also probe deeper into additional factors that impinge upon the mechanisms of knowledge and technology transfer and spillovers. Those could include institutional factors with spatial heterogeneity such as the availability of localised knowledge and technology, of human capital (if that is assumed not to be mobile without friction) and scientific infrastructure, of physical infrastructure, the quality of the regional host innovation system, and the existence of agglomeration (dis-)advantages (clusters, spatial congestion) etc. Additionally, such an extension would allow the analysis to find an answer for the above unresolved question (in Chapter 3) as to whether or not the recent fall in the technology-related motive of FDI into the region is in fact rooted in a lack of quality of the national innovation system in the host economies. Whilst such an extension can be easily and (in Germany) legally done (and has been done in the case of East Germany by the owner of this subset of the database, see for example Günther et al., 2008a and Gauselmann et al., 2011b), it still depends on information about the specific locations of each foreign affiliate – information that is not provided in the other country-specific subsets of the database (again for confidentiality reasons).

Finally, a very important extension of the database would concern the time dimension. So far, very little information is available in the IWH FDI Micro Database to conduct more dynamic analyses. For many issues analysed and tested in this work, the time dimension would be important for an attempt to deepen the current research on the technological effects of inward FDI. Such an extension would allow an analysis based on truly dynamic concepts and theories to be conducted, such as learning at the firm-level (including the question of for example how motives, mandates, product lines, corporate governance structures, intensities of embeddedness etc. have evolved over time in the foreign affiliate) and at the government and local administration level (whether policy reforms have been induced by the corporate sector and how such reforms have impacted on the conditions for technology transfer and spillovers) and finally at the product level (the product lifecycle). The analysis has so far mostly used a given, random point of time, to assess the relevance of conditions for inward manufacturing FDI. A panel structure in the database over an extended period of time would allow the conditions to be tested again, this time taking into consideration changes over time within the same firms.

Many other possible extensions of the database are potentially important. Those listed above, however, are deduced from the most obvious gaps that are evident in the analysis so far by use of the IWH FDI Micro Database.

8.2.2 Future research foci of importance for the technological role of FDI

The most important future research focus that would further develop the research programme is the extension of analysis of technological effects of FDI by looking at the effects of outward FDI originating from CEECs. This extension had already been attempted in the IWH FDI Micro Database from 2008 for East Germany and from 2008/2009 for the

Czech Republic, Hungary, Poland, Romania, and Slovenia (the database used in Chapter 3). Analysis of this data is being prepared currently, yet many important issues remain still unexploited: are the mechanisms described in the literature on the technological effects of outward FDI on the home country statistically significant in CEECs? This research agenda builds on the insights to be gained from the large body of literature on the "pre-selection *vs* learning by exporting" and could extend this to "learning by outward FDI" (see the ongoing research produced under the auspices of the EU 6th Framework Programme of MicroDyn: 'Micro-Dynamics: The competitiveness of firms, regions and industries in the knowledge-based economy'). Successful outward FDI necessitates and at the same time supports the inward and outward internationalisation of R&D and innovation, that is technological activity in the home country. CEECs gradually establish platforms for outward FDI that build on ownership advantages which tend to be mostly in the form of technological advantages.

Another possible path of future development of the research agenda on the technological and developmental effects of FDI could extend the analysis to macro-economic effects. The research programme presented here so far remained at a micro-level to generate empirically tested insights at the firm and institutional levels. This served as a 'micro-foundation' of the overarching research question about the role of FDI projects for technological development in this region of emerging markets. The overarching research question, however, is at a macro-level. It goes without saying that the macro-sphere contains many more developmental issues that form important conditions of economic catching up – the aggregation of all micro-level generated insights will still not account for all technological effects that FDI may have (the idea of holism). The integration of micro and macro foci into one consistent research programme may be the most challenging path of a future research agenda on this overarching objective. Here, the problems discussed in Chapter 1 of this work become relevant: how do the micro-effects, if positively tested empirically, transcend to the macro-level? What is the relationship between macro-economic mechanisms and the sum of micro effects? This includes for example implications of micro-effects for foreign trade specialisation, for capital/labour ratios, for labour markets, for aggregate demand, wages, savings etc., and, vice versa, the implications of those macro-effects for the micro-sphere. The monetary sphere may be a case in point: it plays a key role in providing credit to investors for new capital and R&D, which eventually may lead to technical advances in the economy. Here, the stability of the currency is important inasmuch as it determines the size of interest rates; the depth of the banking sector and the capitalisation of direct capital markets are also important inasmuch as they are the domestic actors granting credits to investors. Whilst the implications of the monetary sphere on technological catch-up development are hence important, their technological effects are even less direct than the 'micro-foundation' at the level of the investor and its investment already are. Further complicating matters, the monetary sphere is itself determined by what happens at the firm level. Here, computable general equilibrium models may provide a starting point for such an analysis; at the very least the propositions to be derived from equilibrium models could serve as a reference point not only to guide the use of empirical data at micro- and macro-levels but also to develop testable hypothesis on how micro-effects transmit to the macro-sphere and back again. As much as this sounds like a formidable way forward, the use of such equilibrium models would however have to reduce heterogeneity and abstract from much of dynamic momentum that drives the technological process to be manageable and able to function as a model. It is the latter concern that led the research on technological development to split up into two families, one of a model-based type and the other devoid of such a straitjacket and yet much more able to account for the dynamic nature of technical advance. It may indeed never be possible to fully amalgamate those two strands of research, and yet each of the two may add their own important insights into the common overarching research question. The current trend of adding an increasing number of macro-issues to an originally micro-level analysis, and the attempts to generate more realistic micro-foundations to originally macro-level research may prove to be the most fruitful path forward. Hereby, institutions may serve as a useful link.

Finally, a future political economy approach could derive policy-relevant implications of analyses much along the lines discussed above and test whether the policy predicaments are helpful and productive or even efficient (in view of an alternative use of resources). It is in particular the latter issue that would again require a framework of a computable general equilibrium model, if the effects of different policy instruments are to be quantified.[5] A particular related problem is the question of coherency of a proposed set of instruments (policy mix).

In general, the motivation of policy interventions is based first and foremost on their expected capability to empower markets to fully reap dynamic capabilities of the economy by way of learning-by-doing (the learning curve), second on (the perceived existence of) market failure or externalities or gaps between private and social returns, and third on the necessity to keep up with international competition on locational conditions (see for example the signalling effect for FDI).[6]

8.2.3 Extensions to the use of theories or concepts

The theoretical or conceptual basis of the analysis presented in this work has already drawn on many different bodies of literature. Still, other theories and concepts may be usefully integrated into the research programme and may add important additional, because they are complementary, foci for the analysis of technological and developmental effects of FDI. Regional or spatial economics, for example, holds important insights into advantages and costs of agglomerations, spatial density in economic activity. The obvious link to the research presented in this work would be for example the embeddedness issue. Along much the same lines, network analysis could add important insights, if applied to the internal and external Cantwell type of networks within and around TNCs. This could tie into the analysis of corporate governance (autonomy, size of equity shares etc.) and into the analysis of regional and sectoral clusters in which FDIs form part. This is what the mode 2 (see for example Gibbons et al., 1994), triple helix (see for example Etzkowitz and Leydesdorff, 2000; Leydesdorff, 2004; Etzkowitz, 2008), and innovation systems approaches (see for example Freeman, 1987; Lundvall, 1992; Nelson, 1993; Patel and Pavitt, 1994; and Metcalfe, 1995) have attempted to do. The network-alignment approach of von Tunzelmann, with its focus on the identification of network misalignments, may offer a valuable method to analyse the capabilities of the various actors in innovation systems and the ability of nodes within the emerging networks to function.

Whatever direction such extension may take, any such research has to duly account for heterogeneity amongst investors and amongst domestic firms, and heterogeneity between different host and home countries. The theoretical approach most suited to this is probably Cantwell's concept of technological accumulation (for an example, see Jindra, 2011). It assumes that each foreign affiliate and each foreign affiliate in the host economy (in fact each actor in the host innovation system) may be different, and that such differences drive the extent, direction, and magnitude of effect on knowledge and technology diffusion. Furthermore, any research agenda on this issue has to develop a coherent conceptual or theoretical basis embracing the most important issues related to technological development at the firm level with the macro- and institutional levels. Only a sound conceptual or theoretical basis at all those levels can further enrich the ongoing discussion about the conditions of technological catch-up development and what economic policy can do to improve those conditions to thereby speed up or solidify catching up in emerging markets. This is what the work presented here and the planned future research agenda outlined above is attempting to do.

Annexes

9.1   The questionnaire for the 2006–2007 wave

IWH FDI MICRO DATABASE

Survey 2007

COORDINATOR:	**Dr J. Stephan**
	Halle Institute of Economic Research (IWH)
	Kleine Märkerstr. 8, 06108 Halle
CONTRIBUTORS:	**Institute for Economic Research (Slovenia)**
	University of Zagreb (Croatia)
	Group of Aplied Economists (Romania)
	EMAR Marketing Research (Poland)
CONTACT:	**B. Jindra (IWH)**
	Tel: +49-345-7753-834, Fax: +49-345-7753-69 834
	E-mail: bja@iwh-halle.de

This research has been partially financed by the EU Commission, in Framework Programme 6, Priority 7 on "Citizens and Governance in a knowledge based society", contract nr CIT5-028519. The Community is not responsible for the content of the survey, any use that might be made of data. The consortium of contributors would like to thank for helpful advice and support from Prof. Nick von Tunzelmann (SPRU, Sussex University), Prof. Klaus Meyer (Bath University), Prof. Igor Filatotchev (Cass Business School), Prof. Slavo Radosevic (SSEES, University College London), Invest in Germany (Berlin), and Invest in Poland (Warsaw).

Part A: Information about your foreign investor

A "foreign investor" holds a minimum of 10% of equity of another company abroad. The "Foreign investor network" or "Multinational Enterprise (MNE) group" comprises the "foreign parent enterprise" or "headquarter" and other units (domestic and foreign) of the foreign investor. The following questions are related to your firm as a subsidiary or affiliate of the foreign investor. Some questions also relate to your foreign investor itself. In case there are more than one foreign investors owners in your firm, the questions relate to the largest foreign investor in terms of equity or board members today.

V1   **NACE (4-digit)** (based on most important product in terms of share in total sales)

V2   **Please indicate the year of the entry of your foreign investor into your firm?**

V3   **Please indicate the total share in equity held by your foreign investor.**

V3_1   At initial entry

V3_2   2002

V3_3   Today

Important: For Croatia V3_2 refers to 2003.

V4   **Please indicate the type of foreign investor in your firm.** Please choose one option!

1   Multinational Enterprise Group

2   National Enterprise Group[1]

3   Enterprise (single entity)

4   Foreign individual or family

V5   **Please indicate the home country (HQ location) of your foreign investor.**

Important: ISO 3166 2-digit country codes

---

[1] A national enterprise group is composed of different units in the home country, however, its only foreign unit is your firm.

V6 **Please indicate which of the following types of owners currently hold equity or have voting rights in your firm.** Please tick the appropriate box for each type of owner. Please consider all owners including the foreign investor.

V6_1 Foreign large MNE group(s) (more than 250 employees or 50 mil Euros in turnover)

V6_2 Small and medium-sized foreign firm(s)

V6_3 Foreign financial investor(s) (bank and/or investment fund)

V6_4 Domestic government or entity(-ies) under state control

V6_5 Domestic financial investor(s) (bank and/or investment fund)

V6_6 Domestic manager(s) or employees of your own firm

V6_7 Unnamed shareholders

Codes: 1 yes, 0 no, 9 no answer

Important: Please note that variable V6_7 is not avaialbe for Croatia and Slovenia.

V7 **Please indicate what describes best the initial entry mode of your foreign investor.**

V7_1 Partial/full acquisition of a state owned firm as part of the privatisation process

V7_2 Partial/full acquisition of a domestic privately owned firm

V7_3 Partial/full acquisition from another prior foreign investor

V7_4 Partial/full ownership in/of a completely new enterprise

Codes: 1 = partial, 2 = full, 7 = does not apply

V8   **Please rank the importance each of the following strategic motives pursued by the foreign investor at initial entry and today.** Please fill in all cells.

V8_1   To access a new market or to increase the existing share on your domestic market (at entry)

V8_1h   Today

V8_2   To follow foreign key clients that moved to your country (at entry)

V8_2h   Today

V8_3   To increase efficiency across the foreign owner network (at entry)

V8_3h   Today

V8_4   To access location-bound natural resources

V8_4h   Today

V8_5   To access location-bound knowledge, skills, technology-

V8_5h   Today

Codes: 1 = not important; 2 = little important; 3 = important; 4 = very important; 5 = extremely important, 9 no answer

Part B: Information about your firm

V9   **Please approximate the structure of your sales according to the location of your buyer(s) (in %).** Please fill in all cells that apply, otherwise enter 0.

V9_1   Exports to your foreign investor network (headquarters and other foreign units)-

V9_2   Exports to other foreign buyers ---

V9_3   Sales to other domestic subsidiaries of your foreign investor -------------------

V9_4   Sales to other domestic buyers --

V10 Please approximate the structure of your supplies according to the location of the respective supplier(s) (in %) Please fill in all cells that apply, otherwise enter 0.

V10_1 Imports from your foreign investor network (headquarters and other foreign units)-

V10_2 Imports from other foreign suppliers --

V10_3 Supplies from other domestic subsidiaries of your foreign investor ------------------

V10_4 Supplies from other domestic suppliers --

V11 Please approximate the following general information about your firm

V11_1a	Total number of employees 2002---
V11_1b	Total number of employees 2005---
V11_2a	Number of R&D personnel 2002---
V11_2b	Number of R&D personnel 2005---
V11_3a	Value of total assets (in Euro) 2002-------------------------------------
V11_3b	Value of total assets (in Euro) 2005-------------------------------------
V11_4a	Value of total sales (in Euro y) 2002------------------------------------
V11_4b	Value of total sales (in Euro) 2005-------------------------------------
V11_5a	Share of intermediate inputs/supplies (as % of total sales) 2002
V11_5b	Share of intermediate inputs/supplies (as % of total sales) 2005

Important: *Please note for Croatia V11 refer to 2003 and 2006 respectively.*

V12 Please indicate the magnitude of the changes of the categories below over the last three years. Please provide an answer for each category.

V12_1	Earnings before interest and taxes
V12_2	Share of exports (in total sales)
V12_3	Value added per employee
V12_4	Market share on your most relevant market
V12_5	Competition within foreign investor network

Codes: 1 = considerable reduction, 2 = reduction 3 = no change 4 = increase; 5 = considerable increase, 9 = no answer

V13   **Does your firm (not you foreign investor) control own subsidiaries abroad?** If yes, please indicate the number and the respective location(s).

13a   Number
V13_1   North America
V13_2   European Union - 15
V13_3   New EU-member countries
V13_4   Former Soviet Union
V13_5   Asia
V13_6   South East Europe
V13_7   other locations

Codes: 1 – Yes, 2 – No, 9 – no answer

Part C: THE RELATIONSHIP BETWEEN YOUR FIRM AND THE FOREIGN INVESTOR

V14   **Please indicate to which degree the following business functions are currently undertaken either by your firm or the foreign owner network (HQ/other unit).**

V14_1   Production and operational management
V14_2   Market research and marketing
V14_3   Basic and applied research
V14_4   Product development[2]
V14_5   Process engineering[3]
V14_6   Strategic management and planning
V14_6   Investment projects and finance

Codes: 1– only your firm, 2 – mainly your firm, 3 – mainly foreign investor network, 4 = only foreign network, 9 = no answer

---

[2] **Product development** refers to product innovations,which are new or significantly improved goods or services with respect to their characteristics (technical specifications, components, materials, incorporated software) or intended uses (user-friendliness etc.). The product must be new to your firm not necessarily to the market!

[3] **Process engineering** refers to new or improved production methods (e.g. computer-assisted design) or delivery methods (e.g. bar-coded goods-tracking system.) including changes in techniques, equipment and/or software.

V15 Please indicate the extent of responsibilities transfer from headquarters and/or other units to your firm since entry of the foreign investor in the follwing areas.

V15_1   New geographical markets

V15_2   New products

V15_3   New business functions (refers to business function listed in V14)

Codes: 1 = no transfer, 2 = limited transfer, 3= considerable transfer, 4 = full transfer, 9 = no answer

V16 Please indicate to which extent you expect such a transfer in the future.

V16_1   New geographical markets

V16_2   New products

V16_3   New business functions (refers to business function listed in V14)

Codes: 1 = no transfer, 2 = limited transfer, 3= considerable transfer, 4 = full transfer, 9 = no answer

V17 Please estimate the intensity of internal competition within your foreign investor network/ multinational group (i.e. between your firm and other domestic/foreign units or HQ of your foreign investor) **with regard to the following areas.**

V17_1 Serving markets

V17_2 Particular or new business lines

V17_3 Business functions (see question 14)

Codes: 1= no competition, 2 = weak intensity, 3 = strong intensity, 4 = very strong intensity

Important: Please note that variable V17 is not available for Croatia and Slovenia.

Part D: RESEARCH & DEVELOPMENT (R&D) AND INNOVATION IN YOUR FIRM

V18   **Please indicate whether your firm has undertaken any of the below listed types of innovation over the last three years. If "yes", please indicate the innovation intensity in comparison to your competitors in the relevant market.**

V18_1   Product innovation[4] ---

V18_1a   Product innovation intensity---

V18_2   Process innovation[5] --

V18_2a   Process innovation intensity ---

V18_3   Marketing innovation[6] --

V18_3a   Marketing innovation intensity ---------------------------------------

V18_4   Organisational innovation[7] ---

V18_4a   Organisational innovation intensity----------------------------------

Codes: Innovation type: 1 = Yes, 2= No, 9 = no answer

Innovation intensity: 1 = very low, 2 = below average, 3 = average, 4 = above average, 5 = very high, 7= does not apply, 9 = no answer

V19   **Please approximate the annual expenditures on R&D and innovation** (including external R&D services). Please indicate the total value in Euro or as a share of total sales. If it does not apply, please indicate "0".

V19_1a   2002 (in % of total sales)

V19_1b   2005 (in % of total sales)

V19_2a   2002 (in EURO)

V19_2b   2005 (in EURO)

Important: *For Croatia V19 refers to 2003 and 2006 respectively.*

---

[4] **Product innovation:** new or significantly improved good or service. The product must be new to your firm not necessarily to the market!

[5] **Process innovation:** new or improved production or delivery methods including e.g. changes in techniques, equipment and/or software.

[6] **Marketing innovation:**   significant changes in product design, packaging, product placement, product promotion or pricing etc.

[7] **Organisational innovation:** new organisational method in the firm's business practices, workplace organisation, or external relations etc.

V20 **Please approximate the share of new or significantly improved products in your firm's total sales.** Please enter "0" if it does not apply to your firm.

V20a 2002 (in % of total sales)

V20b 2005 (in % of total sales)

Important: *Please note for CroatiaV11refer to 2003 and 2006 respectively.*

V21 **Please indicate the importance of the below listed sources for R&D and innovation in your firm?**

V21_1a Acquisition and purchase of external knowledge from abroad

V21_1b Acquisition and purchase of external knowledge domestically

V21_2a Cooperation with other units of the MNE-network abroad

V21_2b Cooperation with other units of the MNE-network domestically

V21_3a Cooperation with other firms abroad

V21_3b Cooperation with other firms domestically

V21_4a Cooperation with other organisations abroad

V21_4b Cooperation with other organisations domestically

21_5 Access to public and open information

Important: *21_1a to 21_4b are not available for East Germany (EDE and EDE_west)*

V21_1EDE Acquisition and purchase of external knowledge (for example licences and R&D services)

V21_2EDE Cooperation (for example with other units of the MNE network, other firm or organisations such as research institutes)

Important: *21_1/2EDE are only available for East Germany (EDE and EDE_west)*

Codes: 1 = not important; 2 = little important; 3 = important; 4 = very important; 5 = extremely important, 9 = no answer

V22 **Please evaluate the importance of the following sources of technological knowledge for R&D or innovation in your firm both, at entry of your foreign investor and today.**

V22_1a Existing technology of your MNE group embodied in products you already produce without substantial adjustments (at entry)

V22_1b today

V22_2a R&D carried out on your own (at entry)

V22_2b today

V22_3a R&D carried out at the headquarters of your foreign investor network (at entry)

V22_3b today

V22_4a R&D carried out by another unit of foreign investor network (at entry)

V22_4b today

V22_5a R&D carried out in collaboration with suppliers abroad (at entry)

V22_5b today

V22_6a R&D carried out in collaboration with local suppliers (at entry)

V22_6b today

V22_7a R&D carried out in collaboration with customers abroad (at entry)

V22_7b today

V22_8a R&D carried out in collaboration with local customers (at entry)

V22_8b today

V22_9a R&D carried out in collaboration with competitors (strategic alliance) (at entry)

V22_9b today

V22_10a R&D carried out in collaboration with scientific institutions abroad (at entry)

V22_10b today

V22_11a R&D carried out in collaboration with local scientific institutions (at entry)

V22_11b today

Codes: 1 = not important; 2 = little important; 3 = important; 4 = very important; 5 = extremely important, 9 no answer

Important: *Variables V22_5 to V22_9 are __not__ available for the Slovenian and Croatian dataset. In the East German dataset (EDE and EDE_west) „domestic" or „local" corresponds to East Germany only.*

V23 Please evaluate the importance of your own firm as a source of technological knowledge for R&D or innovation for others both, at entry of the foreign investor and today.

V23_1a	Headquarters of your MNE group
V23_1b	today
V23_2a	Other units or subsidiaries of your MNE group
V23_2b	today
V23_3a	Your suppliers abroad
V23_3b	today
V23_4a	Your local suppliers
V23_4b	today
V23_5a	Your customers abroad
V23_5b	today
V23_6a	Your local customers
V23_6b	today
V23_7a	Your competitors abroad
V23_7b	today
V23_8a	Your local competitors
V23_8b	today

Codes: 1 = not important; 2 = little important; 3 = important; 4 = very important; 5 = extremely important, 9 = no answer

Important: *In the Slovenian and Croatian dataset the values for customers and suppliers are identical (V23_3a/b = V23_5a/b, 23_4a/b = 23_6a/b). In the East German dataset (EDE and EDE_west) „local" corresponds to East Germany only. In addition V23_7a/b "abroad" refers to foreign and West German competitors.*

9.2 The questionnaire for the 2008–2009 wave

IWH-FDI-Micro-Database

Questionnaire
Survey 2009

In Hungary, Czech Republic, Poland, Romania and Slovakia

1. Round Services

1. Round Manufacturing

Halle Institute for Economic Research (IWH)
Kleine Märkerstraße 8
06108 Halle (Saale) · Germany

Contact: A. Gauselmann, B. Jindra and P. Marek
Research area „Knowledge and Innovation"
Phone: +49-(0) 345-7753-834
Fax: +49-345-7753-779
E-mail: bja@iwh-halle.de

We would like to thank for support and valuable advice by the Zentrum für Sozialforschung (zsh) Halle e.V., Germany Trade and Invest (Berlin), Dr. Johannes Stephan (TU Freiberg), and Prof. Mark Knell (NIFU-STEP) as IWH-Research Professor. The IWH remains responsible for the content of the survey.

I. Part of the questionnaire

Note: The first part of the survey deals with locational factors and ownership structure of your enterprise.

1. Please let me now your position in your enterprise.

Position

2. Please evaluate the quantitative labour supply at your enterprise's location currently. Please choose between very good, good, poor, very poor

	Quality
Supply with unskilled labour	
Supply with skilled labour	
Supply with apprentices and trainees	
Supply with junior employees with university degree	

3. Please evaluate the supply with government grants and subsidies at your enterprise's location currently. Please choose between very good, good, poor, very poor

	Quality
Availability of investment incentives (government grants or tax incentives)	
Availability of fiscal incentives concerning research & development and innovation	

4. Please evaluate the potential for technological cooperation with the following partners at your enterprise's location currently. Please choose between very good, good, poor, very poor

	Quality
Cooperation potential with universities and other public research institutes	
Cooperation potential with other enterprises (customers, suppliers, competitors)	

5. Please evaluate the socio-cultural surrounding at your enterprise's location currently. Please choose between very good, good, poor, very poor

	Quality
Availability of local cultural activities	
Availability of local health care	
Availability of local housing	
Personal safety of expatriates and foreign personnel	
Availability of child care	
General image of the region	

6. Does your enterprise have one or more foreign investor(s)?

Note: A foreign investor is either a direct shareholder with a minimum of 10 percent equity in your enterprise or constitutes the ultimate owner of your enterprise with a minimum of 25 percent indirect ownership. Foreign shareholders are not limited to enterprise groups, but also include physical persons, foundations, financial investors located abroad.

	Yes, one	Yes, more than one	No
Foreign investor			

⇨ **If you answered „No", please continue with question 13.**

Note: The following questions deal with your foreign investor. In case your enterprise has more than one foreign investor, the following questions refer to the investor, who holds the most shares or voting-rights in your enterprise today.

7. Please indicate the type of foreign investor in your enterprise. Please choose one option.

Note for the Interviewer: A multinational enterprise group is composed of different units in your country, the home country and has at least one affiliate in one more country. A national enterprise group is composed of different units in the home country, however, its only foreign unit is your enterprise. Financial investors include banks, investment and venture capital fonds.

Multinational enterprise group	
National enterprise group	
Enterprise (single entity)	
Individual or family	
Financial investor	
Others	

8. Please indicate the home country (HQ location) of your foreign investor.

	Country
HQ location of foreign investor	

9. Please indicate the year of entry of your foreign investor into your enterprise.

	Year
Entry of foreign investor	

10. Please indicate what describes best the initial entry mode of your foreign investor.

	Yes	No
New foundation of a legally independent enterprise		
Partial acquisition of a legally independent and already existing enterprise		
Acquisition of a legally independent and already existing enterprise by the majority		

11. From your point of view: How important were the following strategic motives for your foreign investor's decision to invest in your enterprise? Please choose between not at all important, not important, important and very important.

	Importance
Access to a foreign market	
Use of cost advantages related to labour, capital, or land	
Use of economies of scale (to produce lager amounts of the same product)	
Use of economies of scope (to implement product differentiation)	
Access to location-bound knowledge and technology	
Access to location-bound natural resources	

12. Please indicate whether the following business functions are currently undertaken either by your enterprise or by your foreign investor. Please choose between: <u>only</u> by your enterprise, <u>mainly</u> by your enterprise, <u>mainly</u> by your foreign investor or <u>only</u> by your foreign investor (*Options on enquiry: <u>Neither</u> by your enterprise <u>nor</u> by your foreign investor and By your enterprise <u>and</u> your foreign investor <u>in equal parts</u>*) .

Note for the interviewer: <u>Strategic Management</u> refers to development, planning and implementation of your enterprise's aims and orientation. The planning interval for strategic management covers usually two to five years. <u>Operational Management</u> (or short-term planning) includes activities geared towards the day-to-day operations of the company. <u>Marketing</u> entails not only advertisement activities but also all activities within the company which aim at increasing the demand for the product (e.g. search for markets, changes to the product according to the preferences of the customers, etc.). <u>Aquisition and Supply</u> includes all input factors required for the production of your enterprise's goods and services. <u>Research and Development</u> refers to experimental development, i.e. systematic creative work to broaden existing knowledge, to gain general applicable insights as well as the use of such knowledge for development of new products, services, and processes.

	Only by your enterprise	Mainly by your enterprise	Mainly by your foreign investor	Only by your foreign investor	*Neither by your enterprise nor by your foreign investor*	*By your enterprise and your foreign investor in equal parts*
Strategic Management						
Operational Management						
Marketing						
Acquisition and Supply						
Research and Development						

II. Part of the Questionnaire

Note: The second part of the survey deals with your enterprise's investment abroad.

13. Did your enterprise undertake foreign direct investment?

Note to the interviewer: Foreign direct investment involves the acquisition of a minimum of 10 per cent of equity in an existing legally independent enterprise abroad or the creation of a legally independent subsidiary/affiliate under your own control abroad. We do not refer to foreign direct investment undertaken by any of your own foreign investor.

	Foreign direct investment
Yes	
No	

⇨ If you answered „No", please continue with question 21.

14. Please indicate the year in which your enterprise undertook the first foreign direct investment.

	Year
First foreign direct investment	

15. Please indicate the type of foreign direct investment undertaken by your enterprise.

	Yes	No
New foundation of a legally independent affiliate		
Initial investment in a legally independent and already existing enterprise		
Acquisition of a legally independent and already existing enterprise by the majority		
Follow-up investment in already existing affiliates of the enterprise		

16. How important were the following strategic motives for your enterprise to undertake foreign direct investment? Please choose: not at all important, not important, important, very important.

Access to a foreign market	
Use of cost advantages related to labour, capital, or land	
Use of economies of scale	
Use of economies of scope	
Access to location-bound knowledge and technology	
Access to location-bound natural resources	

17. Regarding foreign affiliates established by your enterprise: Do they work at an upstream, downstream or the same stage as your enterprise in the production process of your final product or service?

Note for the interviewer: A foreign affiliate at an *upstream stage* could for example be a supplier of raw material, intermediate input, or service for the production of the final product or service at your enterprise. A foreign affiliate at a *downstream stage* could be responsible for the sale or distribution of the final product or service produced by your enterprise. It could also operate in an industry that uses your final product or service as an intermediate input. A foreign affiliate at the *same stage* in the production process produces a final product or delivers a service in the same way as your own enterprise just for a different market.

	Yes	No
Foreign affiliate working at an upstream stage		
Foreign affiliate working at a downstream stage		
Foreign affiliate working at the same production stage		

⇨ **If you answered „No", please continue with question 22.**

18. Please list the country(ies) in which foreign affiliates working at an <u>upstream</u> stage are located.

	Country(ies)
Foreign affiliate working at an upstream stage	

19. Please list the country(ies) in which such foreign affiliates working at an <u>downstream</u> stage are located.

	Country(ies)
Foreign affiliate working at an downstream stage	

20. Please list the country(ies) in which such foreign affiliates working at <u>the same</u> production stage are located.

	Country(ies)
Foreign affiliate working at the same production stage	

III. Part of the questionnaire

The following part of the survey deals with research and development in your enterprise. Research and development refers to experimental development to gain general applicable insights as well as the use of such knowledge for development of new products, services, and processes.

21. Did your enterprise undertake any own research and development (R&D) or did it issue any contracts to external research and development providers in the period from 2007 to 2009?

Research and development	2007 until 2009
Yes	
No	

⇨ **If you answered „No", please continue with question 27.**

22. How many of your enterprise's employees work in the area of R&D currently?

<u>Note:</u> The number of employees entails full and part time employees, however, no internships, leasing workers or temporary personnel.

	Currently
Number of R&D employees	

⇨ If you answered „no" to question 6, please continue with question 24.

23. Which impact did the strategic behavior of your foreign investor have on the number of R&D employees? Did the number of R&D employees...

increase	
decline	
Or did it have no direct effect	

⇨ If you answered „no" to question 13, please continue with question 27.

24. Which impact did your enterprise's foreign direct investment have on the number of R&D employees? Did the number of R&D employees...

increase	
decline	
Or did it have no direct effect	

25. Did your enterprise participate in any R&D co-operation with other enterprises or organizations in the period from 2007 to 2009?

Note: R&D cooperation does involve an active participation of your enterprise in projects jointly undertaken for example with related units of your enterprise group, other non-affiliated enterprises, or non-commercial institutions. This does not imply that participating parties extract an economic value from this cooperation. Pure contracts without any active participation of your enterprise are not considered as cooperation.

R&D cooperation	2007 until 2009
Yes	
No	

If you answered „No", please continue with question 27.

26. With which of the following partners did your enterprise co-operate in the area of R&D? More multiple choice possible

	Yes	No
Your headquarter or own enterprise group		
Local suppliers not part of your enterprise group		
Foreign suppliers not part of your enterprise group		
Local customers not part of your enterprise group		
Foreign customers not part of your enterprise group		
Local research institutions		
Foreign research institutions		

IV. Part oft he questionnaire

Note: This part of the survey deals with innovation. Innovations should be new to your enterprise, not necessarily to the market. A product innovation is the introduction of a good or service that is new or significantly improved with respect to its characteristics or intended uses. This includes significant improvements in technical specifications, components and materials, incorporated software, user friendliness, or other functional characteristics. The innovation can be undertaken by your enterprise alone or in cooperation with any other partner.

27. Did your enterprise implement any product innovation(s) in the period from 2007 to 2009?

Note for the interviewer: *Purely aesthetic modifications of products (such as colour, style, and packaging) are no product innovation. The pure sale of an innovation that was neither developed nor produced in your enterprise does not constitute a product innovation*

Product innovation(s)	2007 until 2009
Yes	
No	

⇨ If you answered „No", please continue with question 31.

28. Please approximate the share of new or significantly improved products in your enterprise's total sales in 2009 until now.

	2009
Share of new or significantly improved products in total sales (in %)	

⇨ If you answered „no" to question 6, please continue with question 30.

29. Which impact did the strategic behavior of your foreign investor have on the share of improved products in total sales in your enterprise? Did the share of improved products in total sales...

increase	
decline	
Or did it have no direct effect	

⇨ If you answered „no" to question 13, please continue with question 31.

30. Which impact did your enterprise's foreign direct investment have on the share of improved products in total sales of your enterprise? Did the share of improved products in total sales...

increase	
decline	
Or did it have no direct effect	

V. Part of the questionnaire

Note: The last part of the survey deals with selected general key figures of your enterprise.

31. How many employees do currently work in your enterprise?

Note: The number of employees entails full and part time employees, however, no internships, leasing workers or temporary personnel.

Note to the interviewer: Please indicate the number of all persons defined above as headcount.

	Currently
Number of employees	

⇨ If you answered „no" to question 6, please continue with question 33.

32. Which impact did the strategic behavior of your foreign investor have on the number of employees in your enterprise? Did the number of employees...

increase	
decline	
Or did it have no direct effect	

⇨ If you answered „no" to question 13, please continue with question 34.

33. Which impact did your enterprise's foreign direct investment have on the number of employees in your enterprise? Did the number of employees...

increase	
decline	
Or did it have no direct effect	

34. Please indicate the share of employees in your enterprise who carry out tasks for which an university degree is required.

	Share (in %)
Employees carrying out tasks requiring an university degree	

⇨ If you answered „no" to question 6, please continue with question 36.

35. Which impact did the strategic behavior of your foreign investor have on the total turnover of your enterprise? Did the total turnover...

increase	
decline	
Or did it have no direct effect	

⇨ If you answered „no" to question 13, please continue with question 37.

36. Which impact did your enterprise's foreign direct investment have on the number of employees in your enterprise? Did the total number of employees...

increase	
decline	
Or did it have no direct effect	

37. Please estimate the share of exports on the total sales of your enterprise in 2009.

	2009
Share of exports on total sales (in %)	

38. Please estimate the share of imports on total intermediate inputs in your enterprise in 2009.

	2009
Share of imports on total intermediate inputs (in %)	

Notes

1 Introduction

1. In methodological terms, the theoretical concept of entrepreneurship is not developed in this analysis; entrepreneurial spirits are rather assumed to be given, and may be related to Schumpeterian entrepreneurship (1911, 1949, Chapter 2) (with reference to its effects, creative destruction and innovations), or Keynes' 'animal spirits (1936, pp. 161–62) (with reference to the sources of entrepreneurial behaviour).

2. Many more important rationales for FDI as drivers of the transition processes are conceivable (see for example Cass, 2007, on "[t]he development of the tax system" (ibid., p. 81). Arguably, this is already a rationale only indirectly associated with inward FDI.

3. Institutions governed by the profit incentive of private investors did in fact exist in the real existing socialisms of most CEE countries; even if negligible: they do not pertain only to very small-scale retail "companies" often owned by a single person, but more interestingly to so-called producers' cooperatives. Rather than serving as "schools of socialism" (Hartwig, 1985), they fuelled the shadow economy and contradicted the system of the central planning (Brezinski, 2000).

4. Competition can be seen as the main driving force for the process of discovery of knowledge, that is, new combinations (production processes) and products and services (see Hayek, 1948 and 1968/1978; Schumpeter, 1912 and 1942).

5. In Hayek's view, a socialist planned system is possible in theory but difficult in practice. This is because (tacit) knowledge is typically decentralised, and all sorts of incentives with respect to knowledge are weak in a socialist planned economy due to a lack of competitive pressure. Incentives can only come from management and by order, and yet the mechanisms of planning prevent rational managers from taking the risk of increasing planned effort, if (i) evaluation (and eventually retaliation) is purely on fulfilment of the plan and (ii) the fulfilment of the increased planned efforts depends on deliveries of intermediate inputs outside the control of the manager.

6. See for example the simple Cobb-Douglas model of $Y = AK^\alpha L^\beta$, or in its intensive form $\frac{Y}{L} = A\left(\frac{K}{L}\right)^\alpha$, where output Y grows (when amounts of labour L remain constant) via increases of the amount of capital K and/or improvements in total factor productivity A.

7. In a recent European comparison using employment hours, value added per hour worked and market exchange rates, the Czech Republic reaches the highest level of unit labour costs amongst CEECs (84% of the German level

in 2010) and Hungary the lowest (65%) (Schröder, 2011). Estonia's unit labour costs are surprisingly high (83%), also Latvia's (82%), whilst Slovakia (72%), Poland (71%), and Lithuania (65%) still have large cost advantages for labour-intensive production. The survey does not calculate unit labour costs for countries like Romania and Bulgaria and the Balkans: here, data on hourly productivity is either not available or not comparable to the data in the other countries. It is important to note, moreover, that these figures are national averages that mask very important regional and sectoral differences. Furthermore, (labour) productivities (and also to some extent wage levels) differ markedly between firms, with foreign investors often calculating their own (planned) productivity levels that correspond to the technology they intend to employ in the foreign affiliate and the envisaged employee training schemes. The figures above hence reflect *ex-post* observed labour unit costs, and are of little strategic value for foreign investors.

8. The demand for competitive markets is of particular relevance in CEECs, as in these countries competition as a constituting system criterion had to be newly installed during transition, and today is still of a lower intensity than in West Europe (see for example Emmert et al., 2005; Hölscher and Stephan, 2004).

9. It is important to add here the level of corruption and grey markets in the absence of a reliable rule of law: those may generate economic activity with positive effects on employment and income, and yet their illegality makes economic activity in such a framework attractive only if rents well above average can be earned to remunerate the risks of illegality. Such rents are then often deducted from employee income. This is not an economic environment where efficient economic activity can thrive, and it still attracts (often foreign) investors, even if the wrong ones.

10. A typical numerical and very simple example of a savings gap is the following: the desired growth rate g equals per Harrod–Domar definition s/k where s is the country's rate of savings and k the capital–output ratio. Given a desired growth rate of for example 7 per cent and a capital–output ratio of 3, the necessary savings rate is 21 per cent (7 per cent times 3). A savings gap emerges where the savings rate that can be mobilised domestically remains below the 21 per cent.

11. Here, of course, the economist expects that FDI will predominantly go into sectors where the host economy has comparative advantages (Ricardo). But, because of the long-term character of FDI and the very dynamic nature of the transition and integration processes in CEECs, we may also assume that FDI will alter the pattern of comparative advantages, that is, possibly away from today's revealed comparative advantages to future potential comparative advantages, giving rise to a new position within the international 'division of labour or production'.

12. See for example Fry (1994) for an account of negative economic effects of inward FDI across 11 developing countries, where FDI was accompanied by lower domestic investment and savings, larger current account deficits and lower economic growth rates.

13. In this respect, recall the case of Ford in early 1992 investing in Hungary: here, the investor was granted protection explicitly targeted at the kind of

investment made in the production of commercial vans (this has to be abolished again, but still is an interesting case in point, see Hölscher and Stephan, 1997, footnote 1).

14. See for example Solow's neoclassical growth theory, where technological progress is linked to the accumulation of physical capital; Romer, Grossman, Helpman et al.'s new growth theory, where technological advance is endogenous, the result of knowledge spillovers and innovative activity; Krugman's new trade theory, where increasing returns to scale, monopolistic competition and product differentiation explain intra-industry foreign trade and international factor movements (leading to multinational production like for example FDI).

15. See for example Dunning and Lundan (2008), for their concept of developmental impact of FDI; Lundvall (1988) on a systemic view of innovative activity of firms and institutions, the national innovation systems; Evenett (2005) on the relationship between competition and economic development; and Malerba et al. (1997) on an interpretation of Schumpeterian competition.

16. It is interesting to note that this analysis can potentially be made useful in assessing the role of internationalisation in general and exporting in particular (see for example Filatotchev et al., 2008, pp. 1135–6) for technological development at the firm level. This effectively reverses the *explanans* and *explanandum*, and could be termed a 'macro-foundation of micro-economics'.

17. "For countries aiming to catch up, the basic challenge is to learn to master new ways of doing things [and that is innovation.] [...] The innovation in catching up involves bringing in and learning to master ways of doing things that may have been used for some time in the advanced economies of the world, even though they are new for the country or region catching up. [...] The record is clear that there is considerable learning that needs to be done to enable the new modes of operation to be got under effective control, and a high chance of failure. These are the hallmarks of innovation, at least in evolutionary economic theory." (Nelson, 2008, p. 15).

18. The PRODGAP project ("Determinants of the productivity gap between EU and CEECs") was coordinated and led by the author of this book. The EU Fifth Framework Programme RTD research project focused on the determinants of productivity gaps between the EU and its accession countries of Estonia, Poland, the Czech and Slovak Republics, Hungary, and Slovenia. The project comprised over 30 researchers in 12 participating institutions, ranging from universities and research institutes to statistical offices. The project ran for three years and was allocated a total budget of approximately €1,200,000.

19. The UKNOW project ("Understanding the Relationship between Knowledge and Competitiveness in the Enlarging European Union") was coordinated and led by a team consisting of J. Günther, B. Jindra, and the author of this book (all at that time at the IWH). The EU Sixth Framework Programme STREP project was concerned with the factors giving rise to the public and private nature of knowledge, and the question as to how those factors shape new understandings of knowledge itself, and hinder or foster knowledge creation, use, and dissemination. The project ran for three years and was allocated a total budget of approximately €2,100,000.

20. As at end 2008, 78.2% of all inward FDI into the ten new member states originated from the 15 old EU members (Hunya, 2010, p. 41). The largest share of non-EU investors came from the USA (3.7%) and Switzerland (3.2%).

2 The Database Used in the Empirical Analysis

1. The project was coordinated by the author of this text; it ran from 2001 to 2004, and was allocated a total budget of approximately €1,200,000.
2. The project was coordinated by a team of three German researchers involving Björn Jindra, Jutta Günther and the author of this text. It ran from 2006 to 2009 and was allocated a total budget of approximately €2,100,000.
3. For further information on the IWH FDI Micro database, seehttp://www.iwh-halle.de/projects/2010/fdi/d/start/asp.
4. The current research potential of the IWH FDI Micro Database is documented by the fact that the data is analysed in cooperation with an international network of Eastern and Western European researchers. The IWH FDI Micro Database also tries to make a potential contribution to the international standardisation and harmonisation of survey activities in the field of economic analysis of MNEs. So far, the research using the IWH FDI Micro Database has generated policy advice at various levels in areas such as investment, R&D and innovation policy. At an international level, research findings have served as input for DG Regional Policy and DG Research of the European Commission, as well as UNCTAD and the OECD.
5. The description of the database in terms of basic population, survey sampling and implementation, representativeness and questionnaires, is taken from IWH (2007, 2009) and from Günther et al. (2011a) with some adjustments to the text.
6. Until 2009 in the case of East Germany, data from the MARKUS database was supplemented by information from the European Investment Monitor, the EU R&D Scoreboard and a list generated by the former Industrial Investment Council. In order to assure a uniform information format between the East German and CEE firms this addition was discontinued in 2010.
7. The chi-square test on the sectoral distribution has to be interpreted carefully, because there are fewer than five observations in three industries. Repeating the test using a higher sectoral aggregation, however, confirmed the results. It seems reasonable to assume that the sectoral distribution of the sample does not differ significantly from the total population.
8. From the level of federal states (NUTS 1) the next level down of disaggregation is 'Regierungsbezirke' (NUTS 2). However, this is a purely administrative unit. The level below that is 'Kreise' (NUTS 3). However, this level results in too few observations to assess representativeness. Between NUTS 2 and NUTS 3 there are 23 'Raumordnungsregionen' (ROR) within East Germany. They are constructed as administrative-functional units that take into account the commuting movements of workers between residence and work. Each *Raumordnungsregion* consist of two to six counties ('Kreise').
9. The questionnaire for East German enterprises has three additional questions. Since the principal content is the same for both questionnaires, this differentiation is omitted in the following description.

10. The questionnaire for East German enterprises also has three additional questions, and note 9 also applies here.

3 Foreign Direct Investment Motives and the Match with Locational Conditions in Central East Europe

The analysis in this chapter is to a large extent drawn from a previous publication, 'What drives FDI into Central East Europe? Evidence from the IWH-FDI-Micro Database' (with A. Gauselmann and M. Knell), *Post-Communist Economies*, Vol. 23, No. 3, pp. 343–58 (2011a). The text has been modified to suit the purpose and structure of this work, whilst most of the empirical analysis remained unchanged. The idea and conceptualisation of the analysis was generated collaboratively between the three authors, while most of the empirical analysis was conducted by A. Gauselmann at the IWH, the owner of the original data.

1. It is important to reiterate (see Chapter 1) that *ex-post* unit labour cost advantages are of a lesser strategic importance for foreign investors, because unit labour costs are to an important extent a variable that the foreign investor can control according to the kind of technology installed in the foreign affiliate and the kind of employee training schemes envisaged. This is markedly different for most other locational determinants, over which foreign investors have less immediate control (even where they attempt to improve the kind of locational conditions that are relevant to them, as explicitly assumed in the theory of technological accumulation, as in Cantwell, 1993).

2. In a related analysis of foreign investments into East Germany, several statistically significant differences between home countries (here distinguishing also between West European countries, emerging markets and non-European developed countries, but also West German investors) turned out to be significant (see Gauselmann and Jindra, 2010).

3. Investment projects originating before 1989 are not considered in the analysis. This excludes four investment projects that originated in 1927, 1932, 1975, and 1980. In four other cases, foreign affiliates (asked in 2008) were expecting foreign investment to materialise in 2009.

4. Between 2000 and 2010, unit labour costs in the manufacturing industry (on a €-basis, because this is the relevant figure for European foreign investors) dropped only in Poland (by on average 3.5% per year), whilst growing in the Czech Republic (3.3%) Slovakia (1.7%), and Hungary (0.8%) (calculated from data in Schröder, 2011; there is no data on Romania). Other sources report falling unit labour costs in Romania of around 1%–2% per annum on average (Ghizdeanu and Tudorescu, 2007, for the period 2003–2007; Tatiana-Roxana, 2009, for 2000–2006).

5. A summary of results of empirical testing with correlation analysis between some locational factors and corresponding strategic motives for each host economy is offered in the concluding Chapter 8.

6. The partial correlation analysis controls for the level of importance of other strategic motives, because they are assumed to be not strictly independent of each other, whereas the individual indicators of locational conditions are each assessed individually. For a discussion of the method of partial correlation analysis, see the data and method section of Chapter 7.

4 Conditions of Internal Technology Transfer and Spillovers between Foreign Investors and Foreign Affiliates in Central East Europe

The analysis presented in this chapter constitutes an improved version of earlier taxonomies and their empirical representations published in a variety of previous publications, e.g. 'The Potentials for Technology Transfer via Foreign Direct Investment in Central East Europe – Results of a Field Study', *East-West Journal of Economics and Business*, Vol. VIII, Nos 1&2, pp. 19–41 (2005) and 'Control of FDI subsidiaries and reciprocal technology transfer', and in H. Brezinski and B. Leick (eds) (2006), *Kooperationsperspektiven deutscher Unternehmen in östlichen Grenzregionen* (Freiberg: TU Bergakademie Freiberg), pp. 69–86. The taxonomy with its empirical representation was developed by the author of this study; the publications above contain more empirical descriptions and analyses (share of outward processing trade, share of foreign affiliates in exports, etc.), which were carried out mainly by the co-author, Dr J. Hamar.

1. With the foreign affiliate forming the subject of our field study, absorptive capacities are interrogated only at this level. The host economy's absorptive capacity, a possible further determinant of technology transfer, does not form part of this analysis.
2. The issue of internationalisation of R&D is assessed in detail in Chapter 5 on the role of the host innovation system for the foreign affiliate. In this chapter, the issue is treated as a determinant of internal technology diffusion, whereas in Chapter 5 it forms a condition for foreign affiliates' own technological activity.
3. OPT may not even involve ownership, but today constitutes one important "cornerstone of a new global strategy adopted by multinational companies since the late 1980s" (Andreff, 2009, p. 29).
4. In a study by use of the pilot study of the IWH FDI Micro Database in 2002, Hamar and Stephan (2005 and 2006) calculate the share of OPT type of FDI projects in CEECs and find that the Slovak panel contains the largest shares of OPT-type foreign affiliates, followed, at a significant distance, by Hungary, Poland, and Slovenia. The lowest share of foreign affiliates of an OPT type is found in the Estonian panel.
5. Interestingly, von Tunzelmann (1995) reports for the case of Intel that allegedly the company in the US had to re-import production methods from its assembly plant in Taiwan in the later 1980s, because the headquarters had lost such capabilities at home (p. 372).
6. Both those conceptual developments are discussed in more detail in Chapter 5 on the role of the host innovation system for the foreign affiliate, because they were developed in the context of interaction between foreign affiliate and host economy.
7. As just one case in point, 3M (Minnesota Mining and Manufacturing) invested in Canada in 1951 as a sales affiliate. By 1990, the affiliate was operating its own development units and was responsible for the launching of new products (see e.g. Birkinshaw and Hood, 1998, p. 2).
8. This issue reoccurs in Chapter 5, there with a view on the role of mutual trust and social capital as conditions for external technology transfer and spillovers.

9. In the cases of Lithuania and Romania, foreign affiliates did experience faster TFP growth, but only those that were majority-owned.

10. Needless to say, the strategy of the foreign investor with regard to the extent to which it allows its foreign affiliate to tap its technology or knowledge-related 'ownership advantage' is of paramount importance. However, this analysis assumes that even in the most extreme case of technological marginalisation of an affiliate by its investor, the foreign affiliate will be able to benefit from its links to the foreign firm with a more advanced or at the least different, complementary knowledge stock.

11. This corresponds to the distinction of short-term and long-term impacts, as conceptualised by von Tunzelmann (2004) in his 'network-alignment'. Compare this interpretation with e.g. Moran and Bergsten 1998, and for "open networking" or "strategic technology transfer", see Dyker and von Tunzelmann (2001).

12. In Szalavetz (2000), this link between a change of level of autonomy and the slope of the learning curve is conceptualised by distinguishing between static and dynamic modernisation effects of FDI. Here, static modernisation effects root in low autonomy in all but operational functions and lead the foreign affiliate to achieve production capability and similar efficiency levels as in the parent company. Unless the autonomy position of the foreign affiliate is upgraded in the following, affiliate growth (in sales, exports, etc.) remains static. Dynamic effects of affiliate development only set in with the subsidiary assuming responsibility for additional business functions (functional upgrading).

13. Even though this classification scheme is widely used in the empirical literature, there are important difficulties involved: this classification is often at a two-digit NACE level which are not always homogeneous with respect to the technology intensity of all their member industries. In addition, foreign investors in sectors which are typically considered to exhibit a high technology intensity might tend to allocate their less technology-intensive parts of their production chain to host countries with lower wages and productivity levels.

14. Noteworthy, however, is the observation that standard deviations are very high in the case of East German foreign affiliates (and only surpassed by Romanian affiliates), suggesting that the deviations from the average over all foreign affiliates hide significant heterogeneity between firms.

15. Important to note, however, that publicly funded universities in CEE tend to be less dogmatic and hence able to adjust to the reforms quite quickly (see e.g. Schumann, 2006, for the case of Estonia). Even if this is true, the effects of this on the ability of these actors in the national innovation systems will take time to feed through the systems, as the better educated people tend to be rather young (see e.g. Welfens and Borbély, 2009, p. 131).

16. Negative and positive deviations even out in the picture for all foreign affiliates; this is per definition necessarily so, and it is a result of the particular construction of the composite indicator with equal weights for all firms. At the same time, it is clear that this does not apply to country subsets. Here, a right-hand bias indicates comparatively lower absorptive capacities, and a left-hand bias higher capacities.

5 Central and East European Innovation Systems as Knowledge Sources for Foreign Affiliates' Own Technological Activity in CEE

The ideas and methods of analysis in this chapter draw partially on a number of previous publications: "'FDI and the National Innovation System – Evidence from Central and Eastern Europe', in D. Dyker (ed.) (2011), *Network Dynamics in Emerging Regions of Europe* (Imperial College Press), pp. 303–332 (with J. Günther and B. Jindra)", and "'Does Local Technology matter for Foreign Investors in Central and Eastern Europe? – Evidence from the IWH FDI Micro Database' (with J. Günther and B. Jindra), Special Issue of the *Journal of East West Business on Market Entry and Operational Strategies of MNEs in Transition and Emerging Economies*, Vol. 15, No. 3&4, pp. 210–247 (2009)". The text and empirical analyses presented below are new and original.

1. Of course, not all FDI projects in CEECs by far follow a local knowledge-seeking motive; certainly only a small number treat this motive as dominant (see Chapter 3). It is however an empirical fact that the majority of foreign affiliates in CEECs are engaged in some R&D and do produce some innovations, even if those are, in line with the Oslo Manual (OECD, 2005), not necessarily new to the market but new to the firm. Because this already gives rise to technological improvements in the firm and the economy hosting the firm, this is the correct conceptualisation for this work.
2. See also the original version of the product lifecycle hypothesis (Vernon, 1966) or even the most recent book by Porter (1990).
3. See the lifecycle theory and the idea of a comparatively weaker IPR regime in these countries suggested as an explanation for this in Chapter 1 (1.3.3 The method of analysing conditions of technology transfer and spillovers). The effect of different strengths of IPR regimes is reviewed in Chapter 7.
4. This, however, is assumed to necessarily be accompanied by a high degree of consistency in shared values across the MNC (however, not easily empirically testable). The adverse case would give rise to the risk of 'empire building' in the foreign affiliate, thereby increasing the internal market administration costs (Birkinshaw, 1998) and lowering efficiency of the internal capital market (Mudambi, 1999).
5. Another connotation used in the literature is "subsidiary evolution" (e.g. Birkinshaw and Hood, 1998) and assumes the perspective of dynamic changes in the capabilities of foreign affiliates that are driven by, amongst other things, autonomy of foreign affiliates and the quality of parent–subsidiary relationships in general.
6. This corresponds to the three drivers of subsidiary development in Birkinshaw and Hood (1997, pp. 341–5): parent development, internal development, and host-country development.
7. The systemic innovation literature considers different territorial and sectoral aggregates when describing the characteristics of systems (national, regional, and local innovation systems, sectoral innovation systems, etc.). The individual choices depend on what is assumed to form a common framework for heterogeneous innovators. For the purpose of this work (and with a view on

the focus on spatial proximity below), such distinctions may be disregarded. The connotation in the literature of 'regional innovation systems' is probably closest to the analysis here, because of the selection of actors in this system for the empirical analysis. Following Cooke et al. (1997), regional innovation systems are defined as a network of regionally interacting actors and institutions from the private and public sector that generate, modify, and diffuse new technologies. Still, the text here refers to 'host innovation systems' and thereby refrains from distinguishing between the different conceptualisations of innovation systems.

8. See the internalisation concept with its transaction cost theory, e.g. Williamson (1975, 1981); and for a critique from an evolutionary perspective, see e.g. Pitelis (1998).

9. It is important to note that this is quite distinct from the situation in Russia and the Community of Independent States, where foreign investors not only face much higher risks of adverse political and administrative interference but also higher transaction costs and more risks associated with the role of particular societal groups (oligarchs and the Russian mafia).

10. The latter criterion makes sure that innovative activity of firms include all measures that result in technological advance within the firm as the object of analysis. Technical advance in the market, the country or on a global scale are not what the chapter focuses on.

11. In a related analysis by Günther et al. (2008b), some regional indicators were used to complement the East German section of the IWH FDI Micro Database of 2006–2007. Those indicators are defined at the level of regional aggregation of *'Raumordnungsregionen'* (ROR), of which there are 23 such administrative functional units within East Germany. This regional disaggregation is a good compromise between the locally very disaggregated NUTS 3 regions and the somewhat overlarge NUTS 2 regions. The indicators include human capital stocks (HRSTO); intensity of R&D expenditures by local higher education institutions; the regional knowledge stock, approximated by the intensity of sectoral patent application; and agglomeration effects, measured by the sectoral employment density. Matching firm-specific data with regional indicators is possible at the IWH, because this institute is the original owner of this section of the database. It still remains impossible for the general academic community to do the same for the other country sets of the database. The results of the East Germany-related analysis is only briefly described in this work, because it adds important insights into regional determinants and the problems involved with such an extension to the analysis of this work. The analysis and results are not presented here in full, because their yield would be outside the scope of this work and has already been published.

12. In fact, this difference is particularly small for Croatian and Romanian foreign affiliates (the identical shares for employment and expenditure in Croatia are coincidental), leaving little room – on average – for external R&D. In the Polish, Slovene and East German cases, the share of foreign affiliates with R&D expenditure is noticeably higher than the share of affiliates with explicit R&D personnel, which leaves room for R&D in excess of intra-mural research efforts.

13. The distinction between shares in gross value added and shares in turnover is intended to correct for differences in firm sizes: R&D expenditure shares

in total turnover tend to develop under-proportionally with increasing firm size, i.e. underestimating the intensity amongst groups of larger foreign affiliates.

14. Importantly, the picture does not change much in the strategic motives that prevailed at the time of initial entry: the mean for knowledge base was then slightly lower, down from 2.5 to 2.2 for foreign investors into Poland, also in the case of Croatia from 3.1 to 2.8, and Romania from 3.0 to 2.8. The ranking, however, remained unchanged in those latter cases.

15. It remains of course unclear as to whether domestic suppliers in fact produce their supplies in the host economy or whether those rather represent imports via a domestic router. Likewise unclear is whether domestic customers in fact use or consume the product in the host economy or whether they again act as exporting agents. This problem may somewhat reduce the robustness of interpretation of local supplies and sales as forming part of the host innovation system, yet the respective actors do form part of that system per definition: effects such as knowledge spillovers between the foreign affiliate and the host economy are insensitive to that problem, and the interpretation of intensity of trade relations with the host economy as a condition for a positive developmental role of FDI remains valid.

16. It has to be noted, however, that standard deviations are high, in each case in excess of 100 per cent of the number employed – the data contains wide variation, and averages should therefore be treated with due care.

17. It is possible to include all dummies for all investment motives, because the motives are not self-excluding and foreign affiliates may select several, one, or none of the motives as particularly important.

18. This amounts to shares of 78.6 per cent in East Germany, 71.0 per cent in Slovenia, 79.0 per cent in Poland, 60.2 per cent in Croatia, and 65.8 per cent in Romania.

19. Taking this information alone would raise a question mark for the sustainability of Polish technical advance: insufficient R&D will eventually force innovativeness to peter off. Alas, the Polish economic development as until recently does not show any signs of weakening technological advancement. On the contrary, the structure of exports has improved. See the case made by Lavigne (1999), that the first investors in Poland were of Polish decent (expatriates) and brought with them very little technology but were, rather, seeking swift profits. This, however, changed fundamentally in later years.

6 Foreign Affiliates as Knowledge Source for Host Regional Innovation Systems in CEE

The analysis in this chapter has not been published yet, all ideas have been developed originally, and the analysis **was** carried out by Dr Stephan.

1. As a counter-effect, foreign affiliates typically try to attract skilled workers to effect the opposite direction of labour mobility (see e.g. Sinani and Meyer, 2004, for the case of Estonia; and Lipsey and Sjöholm, 2005, and Blalock and Gertler, 2005, for the case of Indonesia).

2. This method builds on Caves (1974) and Globerman (1979), who used cross-sectional industry-level data for Australia and Canada respectively, to find a positive impact of foreign investors on local firms.

3. Those include: Bulgaria, the Czech Republic, Estonia, Hungary, Poland, Romania, Slovakia, and Slovenia.

4. Smarzynska Javorcik 's (2004a) study of FDI in Lithuania cannot establish positive spillovers within the same industry.

5. The definition of the relevant market is also left to the managers' own discretion, as a determination by way of e.g. NACE three levels may be too broad to include many non-relevant firms, whilst at the same time excluding other firms in related industries outside the definition of the NACE level that are in fact relevant in terms of competition.

6. The analysis uses unweighted averages, because the two individual indicators for product and process innovation are correlated with a Spearman correlations coefficient of 0.4892 with statistical significance. For the econometric tests, the two indicators have been amalgamated to prevent multicollinearity.

7. This solution was found in test runs and discussions with interviewees and renewed test runs, and has proven to be the best proxy possible in the framework of this fieldwork.

8. The theoretical case motivating the two hypotheses, H(2) and H(3), on augmenting and exploiting strategies is not as clearcut as in the cases of the other hypothesis, and will to some extent depend on the industrial sector. Still, the concept is interesting enough to warrant empirical testing in this framework.

9. This does not always show in the descriptive statistics presented below: here, the complete set of information was used to describe the database to its fullest possible extent.

10. Figures 5_4 (Innovative activity of foreign affiliates across countries: intensity of product innovation) and 5_5 (Innovative activity of foreign affiliates across countries: intensity of process innovation) in Chapter 5 present the two individual intensities of innovation in their original scale.

11. Again, it is consistent to include all dummies for all investment motives, as they are not mutually exclusive.

12. The number of observations (n=197) is much lower than in the analyses of the previous chapter. This is due to the fact that some of the new variables considered here and the dependent variables had been not filled out as much by foreign-affiliate managers as had the variables used in the previous chapter. Excluding e.g. the technological-ability proxy would increase the observations to n=277, excluding additionally the expenditure for technological activity yields a number of observations of 519, which corresponds to the numbers of the analyses in the previous analysis.

13. tech_abil was tested in both specifications (_host and _si). The results presented here are for the _host specification, because the subsequent analysis shows that this specification produces more robust results.

14. In the specification using *tech_abil_si*, *size* becomes significant (and negative) for the test on local competitors with a $P > |z|$ of .031 with an overall pseudo R^2 of .132. For local suppliers, specification using *tech_abil_si*, *exploit* becomes significant (and positive) with a $P > |z|$ of .052, and yet *age* turns insignificant with an overall pseudo R^2 of .207.

7 The Role of Intellectual Property Rights for Technology in FDI into CEE

The analysis in this chapter has not been published yet, but will be published in a special issue of the *Journal of the Knowledge Economy* with the title "STI policy for the Enlarging European Economies: The Generation, Use, and Dissemination of Knowledge in the Heterogeneous European Economies" in 2012, edited by Dr Knell and Dr Stephan. The idea to analyse IPR and FDI by use of the IWH FDI Micro Database has already been published in: 'Foreign Direct Investment in Weak Intellectual Property Rights Regimes – the Example of Post-Socialist Economies', *Post-Communist Economies*, Vol. 23, No. 1, pp. 35–53 (2011). This article uses the same database but a different construction of proxies.

1. To complicate matters, the commercial value of knowledge may equally increase with the use by many, as new ideas are added to the original idea, or if a new technology can, when used widely, become an industrial standard. The important question, however, is who owns what knowledge involved in this process, or who is able to profit from the use of that knowledge.
2. A third indicator, the Ginarte/Park index, exclusively depicts laws on the books and not their effective enforcement. See Maskus (2000) for a discussion of construction and properties of this and related indices. All those indices are ordinal numbers on different scales.
3. See the correlation and regression analyses in IPRI (2009, pp. 29–32). In the IWH FDI Micro Database 2006–2007, the pairwise correlation coefficient amounts to 0.979.
4. This may be the case in some developing and emerging markets; see the frequent allegations against for example the Chinese IPR performance and the latest attempts by their government to reassure (potential) foreign investors with the alleged clampdown against IPR fraud that is so intensively covered by the Chinese media.
5. The prolonged and bumpy negotiations between prospective new EU members and the EU Commission that led up to the implementation of the competition law and policy chapter within the *acquis communautaire* is a case in point (see for example Hölscher and Stephan, 2004): many temporary derogations resulted from this process, and some of those are in fact significant (see for example state aid rules, tax holidays etc.).
6. See most prominently the recent property rights disputes in Russia, and the outright expropriation of foreign assets in Venezuela. It is important to note that such interventions, seen as drastic in the West, have a historical root in the system of economic planning: during socialist times, cross-border ownership rights had to rest by default with the host country, to preserve the consistency of the economic plans of the countries involved (see Brezinski, 1978, pp. 172–3).
7. However, the latter does not necessarily include competition policy, which would be the institution of which IPR regimes form part.
8. A survey of nearly 1500 R&D manufacturing laboratories in the United States shows that 51 per cent of innovations were protected by trade secrets and only 35 per cent by patents (Cohen et al., 2000). This promotes the authors of the World Investment Report (UNCTAD, 2005) to assume that "[t]o the

extent that the R&D process involves sensitive information, TNCs will always seek to protect trade secrets against disclosure." (ibid., p. 209).

9. Those motives relate to the strategic motive as of the time of the field study, i.e. 2005, and may assume values between 1 for "not important" to 5 for "extremely important".

10. The use of ordered probit analysis is determined by the fact that the dependent variable *IPR gap* is ranked on an ordinal scale.

11. The OLS regressions are tested in a linear specification and after transformation of dependent and independent variables into natural logs. The use of these methods here are determined by the fact that the dependent variables (*comp.sal* and *comp.sup*) are continuous variables. In the case of the comparison between foreign and domestic science institutions, the ordinal scale requires the use of ordered probit analysis.

8 Conclusions on the Technological Role of FDI into CEE

1. The latter issue is analysed only indirectly in this work, due to the character of the IHW FDI Micro Database that focuses on the foreign affiliates and not directly on the actors of the host innovation system.

2. This may be the opposite for the perspective of internal direct technology and knowledge transfer: here, greenfield investments may actually receive more sensitive technology, if only because the danger of unwanted dissipation of ownership-advantages à *la* Dunning is more easily prevented. Alas, for a sustained, long-lasting, and possibly dynamic positive technological role, this 'first-order' condition remains insufficient, because it explicitly does not include the possibility of other actors in the host innovation system benefiting from this ownership advantage.

3. In fact Croatia and Poland have better rankings as a country than the rankings for their endowments with FDI projects in the IWH FDI Micro Database. This is effected by Romania, where the average gap is much lower than the WEF indicator would suggest. Standard deviation in this country is the highest.

4. In any case, the development of such a database is costly, and the development so far was governed by the varying foci of research project that provided the necessary funding. Future extensions will again depend on the ability to raise funding, and are likely to again be subject to varying topical foci.

5. And yet, sound economic-policy analysis can also produce interesting insights if based on simple regression analysis. See for example Harding and Javorcik (2011) analysing the effect of investment promoting policies across 124 countries by way of panel regression models (and convincingly concluding that investment promotion works in developing countries but not in industrialised economies).

6. Interesting to note that one of the main players in the field of technological development, the Science and Technology Policy Research Institute at the University of Sussex is now propagating, under the new leadership of Mariana Mazzucato, to research into a view beyond market failure as a motivation for policy interventions.

References

Aitken, B.J. and A.E. Harrison (1999) 'Do domestic firms benefit from direct foreign investment? Evidence from Venezuela', *American Economic Review* 89, pp. 605–18.

Aitken, B.J., G.H. Hanson, and A.E. Harrison (1997) 'Spillovers, Foreign Investment, and Export Behavior', *Journal of International Economics* 43/1–2, pp. 103–132.

Alcacer, J. and W. Chung (2007) 'Location Strategies and Knowledge Spillovers', *Management Science* 53/5, pp. 760–76.

Almeida, P. (1996) 'Knowledge sourcing by foreign multinationals: patent citation analysis in the U.S. semiconductor industry', *Strategic Management Journal* 17, pp. 155–65.

Altenburg, T. (2000) *Linkages and Spill-overs between Transnational Corporations and Small and Medium-Sized Enterprises in Developing Countries – Opportunities and Policies*, German Development Institute Report and Working Paper 5/2000, mimeo.

Altomonte, C. (2000) 'Economic determinants and institutional frameworks: FDI in economies in transition', *Transnational Corporations* 9/2, pp. 75–106.

Altomonte, C. and C. Guagliano (2003) 'Comparative Study of FDI in Central and Eastern Europe and the Mediterranean', *Economic Systems* 27/2, pp. 223–46.

Ambos, B. (2005) 'Foreign direct investment in industrial research and development: a study of German MNCs', *Research Policy* 34/4, pp. 395–410.

Andersen, B., F. Rossi, and J. Stephan (2010) 'Property (IP) marketplaces and how they work: Evidence from German pharmaceutical firms', *Intereconomics – Review of European Economic Policy* 45/1, pp. 35–41.

Andersson, U. and M. Forsgren (2000) 'In search of centre of excellence: network embeddedness and subsidiary roles in multinational corporations', *Management International Review* 40/4, pp. 329–50.

Andersson, U., I. Björkman, and M. Forsgren (2005) 'Managing subsidiary knowledge creation: The effect of control mechanisms on subsidiary local embeddedness', *International Business Review* 14, pp. 521–38.

Andersson, U., M. Forsgren, and U. Holm (2002) 'The strategic impact of external networks: subsidiary performance and competence development in the multinational corporation', *Strategic Management Journal* 23/11, pp. 979–96.

Andreff, M. and W. Andreff (2001a) 'Outward-Processing Trade between France and Central and Eastern European Countries: Is There a Substitution to France-Maghreb Outward-Processing Trade?', *Acta Oeconomica* 51/1, pp. 65–106.

Andreff, M. and W. Andreff (2001b) 'Outward-processing trade and foreign direct investment from France into East European countries', in N. Fabry and S. Zeghni (eds), *Transition in Asia and Central and Eastern Europe: A closed door, two open windows* (Huntington: Nova Science Publishers), pp. 117–49.

Andreff, M. and W. Andreff (2005) 'La concurrence pour l'investissement direct étranger entre les nouveaux et les anciens membres de l'Union Européenne élargie' (The competition between new and former European Union members for foreign direct investment), *Economie appliquée* 58/4, pp. 71–106.

Andreff, W. (2009) 'Outsourcing in the new strategy of multinational companies: foreign investment, international subcontracting and production relocation', Papeles de Europa 18, pp. 5–34.

Arora, A. (2009) 'The Economics of Intellectual Property: Suggestions for Further Research in Developing Countries and Countries with Economies in Transition. Intellectual Property Rights and the International Transfer of Technology: Setting Out an Agenda for Empirical Research in Developing Countries', *WIPO*, pp. 41–58.

Arora, A. and A. Gambardella (1990) 'Complementarities and external linkages: The strategies of large firms in biotechnology', *Journal of Industrial Economics* 38/4, pp. 361–79.

Åslund, A. (2002) *Building Capitalism – The Transformation of the Former Soviet Bloc* (New York: Cambridge University Press).

Audretsch, D. and M.P. Feldman (1996) 'R&D Spillovers and the Geography of Innovation and Production', *American Economic Review* 86, pp. 630–40.

Audretsch, D. and M.P. Feldman (2003) 'Knowledge Spillovers and the Geography of Innovation', in J. V. Henderson and J. Thisse (eds) *Handbook of Urban and Regional Economics* 4 (Amsterdam: North Holland Publishing), pp. 2713–39.

Band, J. and M. Katoh (1995) Interfaces on Trial: Intellectual Property and Interoperability in the Global Software Industry (Boulder: Westview Press).

Barrios, S., H. Goerg, and E. Strobl (2003) 'Explaining Firms' Export Behaviour: R&D, Spillovers and the Destination Market', *Oxford Bulletin of Economics and Statistics* 65/4, pp. 475–96.

Barrios, S. and E. Strobl (2002) 'Foreign direct investment and productivity spillovers: Evidence from the Spanish experience', *Review of World Economics* (*Weltwirtschaftliches Archiv*) 138/3, pp. 459–81.

Bartlett, C. A. and S. Ghoshal (1986) 'Tap your subsidiaries for global reach', *Harvard Business Review* 64/6, pp. 87–94.

Bartlett, C. A. and S. Ghoshal (2004) 'Transnational Management', in B. DeWitt and R. Meyer (eds) *Strategy: Process, Content, Context*. 3rd ed., pp. 577–88 (London: Thomson Learning).

Bartlett, C.A. and S. Ghoshal (1989) *Managing Across Borders: The Transnational Solution* (Boston: Harvard Business School Press).

Bartlett, C.A., S. Ghoshal, and J. Birkinshaw (2004) *Transnational management: texts, cases, and readings in Cross-border management* (New York: McGraw Hill/Irwin).

Bascavusoglu, E. and M.P. Zuniga (2002) *Foreign Patent Rights, Technology and Disembodied Knowledge Transfer Cross Borders: An Empirical Application*, University of Paris I, mimeo.

Basile, R. (2004) 'Acquisition versus Greenfield investment: the location of foreign manufacturers in Italy', *Regional Science and Urban Economics* 34/1, pp. 3–25.

Beamisch, P. (1993) 'The characteristics of joint ventures in the People's Republic of China', *Journal of International Marketing* 1, pp. 29–48.

Békés, G., J. Kleinert, and F. Toubal (2009) *Spillovers from Multinationals to Heterogeneous Domestic Firms: Evidence from Hungary,* Document de travail, Centre d'Études Prospectives et d' Informations Internationales, 2009–31, Paris.

Belderbos, R., K. Fukao, and Hyeog Ug Kwon (2006) *Intellectual Property Rights Protection and the Location of Research and Development Activities by Multinational*

Firms, Hi-Stat Discussion Paper Series, Institute of Economic Research, Hitotsubashi University, Tokyo.

Bellak, Ch., M. Leibrecht, and A. Riedl (2008) 'Labour costs and FDI flows into central and Eastern European Countries: A survey of the literature and empirical evidence', *Structural Change and Economic Dynamics* 19, pp. 17–37.

Bevan, A., S. Estrin, and K.E. Meyer (2004) 'Foreign investment location and institutional development in transition economies', *International Business Review* 13, pp. 43–64.

Birkinshaw, J. (1997) 'Entrepreneurship in multinational corporations: the characteristics of subsidiary initiatives', *Strategic Management Journal* 18/3, pp. 207–29.

Birkinshaw, J. (1998) 'Corporate Entrepreneurship in Network Organizations: How Subsidiary Initiative Drives Internal Market Efficiency', *European Management Journal* 16/3, pp. 355–64.

Birkinshaw, J. and J. Ridderstrale (1999) 'Fighting the corporate immune system: a process study of subsidiary initiatives in multinational corporations', *International Business Review* 8, 149–80.

Birkinshaw, J. and N. Hood (1998) 'Multinational Subsidiary Evolution: Capability and Charter Change in Foreign-Owned Subsidiary Companies', *The Academy of Management Review* 23/4, pp. 773–95.

Birkinshaw, J. and N. Hood (eds) (1998) *Multinational Corporate Evolution and Subsidiary Development* (London: Macmillan).

Birkinshaw, J. M. and A. J. Morrison (1996) 'Configurations of strategy and structure in multinational subsidiaries', *Journal of International Business Studies* 26/4, pp. 729–94.

Birkinshaw, J.M. and N. Hood (1997) 'An empirical study of development processes in foreign-owned subsidiaries in Canada and Scotland', *MIR: Management International Review* 37/4, pp. 339–64.

Birkinshaw, J.M., N. Hood, and S. Jonsson (1998) 'Building firm specific advantages in multinational corporations: the role of subsidiary initiative', *Strategic Management Journal* 19/3, pp. 221–41.

Blalock, G. and J. Gertler (2004) *Welfare gains from FDI through technology transfer to local suppliers*, University of California, Berkeley, mimeo.

Blalock, G. and J. Gertler (2005) 'Foreign Direct Investment and Externalities: The Case for Public Intervention', in T.H. Moran, E.M. Graham, and M. Blomström (eds), *Does Foreign Direct Investment promote Development?* (Washington, DC: Institute for International Economics and Center for Global Development), pp. 73–106.

Blanc, H. and C. Sierra (1999) 'The internationalisation of R&D by multinationals: a trade-off between external and internal proximity', *Cambridge Journal of Economics* 23, 187–206.

Blanchard, O. (1997) *The Economics of Post-Communist Transition* (Oxford: Oxford University Press).

Blomström, M. and A. Kokko (1998) 'Multinational Corporations and Spillovers', *Journal of Economic Surveys* 12/2, pp. 1–31.

Blomström, M. and A. Kokko (2003) *Human Capital and inward FDI*, CEPR research network on Foreign Direct Investment and the Multinational Corporation: New Theories and Evidence, Working Paper 167, mimeo.

Blomström, M. and E. Wolff (1994) 'Multinational Corporations and Productivity Convergence in Mexico', in W. Baumol, R. Nelson, and E. Wolf (eds), *Convergence* of Productivity: Cross-National Studies and Historical Evidence (Oxford: Oxford University Press) pp. 115–43.

Blomström, M., S. Globerman, and A. Kokko (2001) 'The determinants of host country spillovers from foreign direct investment: review and synthesis of the literature', in N. Pain (ed.) *Inward Investment, Technological Change and Growth* (Basingstoke: Palgrave) pp. 34–65.

Borensztein, E., J. De Gregorio, and J-W. Lee (1998) 'How Does Foreign Direct Investment Affect Economic Growth?', *Journal of International Economics* 45, pp. 115–35.

Bosco, M.G. (2001) 'Does FDI contribute to technological spillovers and growth? A panel data analysis of Hungarian firms.' *Transnational Corporations* 10/ 1, pp. 43–68.

Boudier-Bensebaa, F. (2005) 'Agglomeration economies and location choice. Foreign direct investment in Hungary', *Economics of Transition* 13/4, 605–28.

Boudier-Bensebaa, F. and H. Brezinski (2000) *Outward-processing trade between Germany and the Central Eastern European countries.* VIth ICCEES World Congress, Tampere, 29.7.–3.8., mimeo.

Braga, C.A.P. and C. Fink (1989) 'The Relationship between Intellectual Property Rights and Foreign Direct Investment', *Duke Journal of Comparative & International Law* 9, pp. 163–87.

Branstetter, L.G., R. Fisman, and C.F. Foley (2005) *Do Stronger Intellectual Property Rights Increase International Technology Transfer? Empirical Evidence from U.S. Firm-Level Panel Data*, Columbia Business School and NBER, mimeo.

Brezinski, H. (1978) *Internationale Wirtschaftsplanung im RGW*, Schriften der Gesamthochschule Paderbrn, Reihe Wirtschaftswissenschaften (Paderborn: Verlag Ferdinand Schöningh).

Brezinski, H. (1991) 'Institutional Framework of Reforms: A Discussion', in M. Kaser and A.M. Vacic (eds) *Reforms in Foreign Economic Relations of Eastern Europe and the Soviet Union*, Economic Studies of the UNECE, No. 2, pp. 45–8.

Brezinski, H. (1992) 'Privatisation in East Germany', *MOCT-MOST* 2/1, pp. 3–21.

Brezinski, H. (2000) 'The Microeconomics of Producers´ Cooperatives in Transition', *Economic Systems* 24/4, pp. 370–74.

Brezinski, H. (ed.) (2006) *Kooperationsperspektiven deutscher Unternehmen in östlichen Grenzregionen* (Freiberg: TU Bergakademie Freiberg).

Brezinski, H. and J. Stephan (2011) 'Capital inflows, Current Accounts, and Exchange Rate Regimes in Central East Europe during and after the Global Financial Crisis', *Jahrbuch für Wirtschaftswissenschaften* 62/1, pp. 22–39.

Brouthers, K.D. and L.E. Brouthers (2001) 'Explaining the National Cultural Distance Paradox', *Journal of International Business Studies* 32/1, pp. 177–89.

Buck, T., I. Filatotchev, M. Wright, and N. Dyomina (2003) 'Insider Ownership, Human Resource Strategies and Performance in a Transition Economy', *Journal of International Business Studies* 34/6, pp. 530–49.

Buckley, P.J. and M. Casson (1976) *The Future of the Multinational Enterprise* (London: Macmillan).

Buckley, P.J. and M. Casson (1996) 'An Economic Model of International Joint Venture Strategy', *Journal of International Business Studies* 27/5, pp. 849–76.

Bundesforschungsanstalt für Landeskunde und Raumordnung (1996) *Neuabgrenzung von Raumordnungsregionen nach den Gebietsreformen in den neuen Bundesländern*, Bonn.

Bureau van Dijk (2010), *Amadeus – A database of comparable financial information for public and private companies across Europe*, Online Brochure, URL: http://www.bvdinfo.com/ getattachment/da04b736-b71a-4c6f-acc6-ba2a9e423bf9/ Amadeus.aspx (Date: 02–11–2011)

Cantwell, J. (1989) *Technological innovations in multinational corporations* (Oxford: Blackwell).

Cantwell, J. (1992) 'The theory of technological competence and its application to international production', in D.G. McFetridge (ed.) *Foreign Investment, Technology and Economic Growth* (Calgary: University of Calgary Press) pp. 33– 67.

Cantwell, J. (1993) 'The internationalization of technological activity and its implications for competitiveness', in O. Granstrand, H. Hakanson, and S. Sjolander (eds) *Technology Management and International Business* (Chichester: Wiley), pp. 137–62.

Cantwell, J. (2000), 'A survey of theories in international production', in C.N. Pitelis and R. Sugden (eds) *The Nature of the Transnational Firm*, 2nd edition (London: Routledge), pp. 10–56.

Cantwell, J. and L. Piscitello (1999) 'The emergence of corporate international networks for the accumulation of dispersed technological competences', *Management International Review* 39, pp. 123–47.

Cantwell, J. and L. Piscitello (2002) 'The Location of Technological Activities of MNCs in the European Regions: The Role of Spillovers and Local Competencies', *Journal of International Management* 8, pp. 69–96.

Cantwell, J. and L. Piscitello (2005) 'Recent Location of Foreign-owned Research and Development Activities by Large Multinational Corporations in the European Regions: The Role of Spillovers and Externalities', *Regional Studies* 39/1, pp. 1–16.

Cantwell, J. and O.E.M. Janne (1999) 'The role of multinational corporations and national states in the globalisation of innovatory capacity', *The European perspective, technology Analysis & Strategic Management* 12/2, pp. 155–72.

Cantwell, J. and R. Mudambi (2005) 'MNE Competence creating subsidiary mandates', *Strategic Management Journal* 26/12, pp. 1109–28.

Cantwell, J. and S. Iammarino (1998) 'MNCs, technological innovation and regional systems in the EU: some evidence in the Italian case', *International Journal of the Economics of Business* 5/3, pp. 383–408.

Cantwell, J. and S. Iammarino (2003) *Multinational Corporations and European Regional Systems of Innovation* (London: Routledge).

Cantwell, J. and Santangelo (2000) 'Capitalism, innovation and profits in the new technoeconomic paradigm', *Journal of Evolutionary Economics* 10/1–2, pp. 131–57.

Cantwell, J., S. Iammarino, and C.A. Noonan (2001) 'Sticky Places in Slippery Space – the Location of Innovation by MNCs in the European Regions', in N. Pain (ed.) *Inward Investment, Technological Change and Growth* (London: Macmillan).

Cass, F. (2007) 'Attracting FDI to Transition Countries: The Use of Incentives and Promotion Agencies', *Transnational Corporations* 16/2, pp. 77–122.

Castellani, D. and A. Zanfei (2003) 'Technology Gaps, Absorptive Capacity and the Impact of Inward Investments on Productivity of European firms', *Economics of Innovation and New Technology* 12/6, pp. 555–76.

Caves, R. (1974) 'Multinational Firms, Competition and Productivity in Host-Country Markets', *Economica* 41/162, pp. 176–93.

Cernat, L. (2006), *Europeanization, Varieties of Capitalism and Economic Performance in Central and Eastern Europe* (New York: Palgrave Macmillan).

Chesbrough, H. (2003) *Open innovation – The new Imperative for Creating and Profiting from technology,* (Boston: Harvard Business School Press).

Chidlow, A., L. Salciuviene, and S. Young (2009) 'Regional determinants of inward FDI distribution in Poland', *International Business Review* 18/2, pp. 119–33.

Child, J. (1993) 'Society and enterprise between hierarchy and market', in J. Child, M. Crozier, and R. Mayntz (eds), *Societal Change between Market and Organization* (Aldershot: Avebury), pp. 203–26.

Chung, W. and J. Alcacer (2002) 'Knowledge Seeking and Location Choice of Foreign Investment in the United States', *Management Science* 48/12, pp. 1535–54.

Claessens, S. and S. Djankov (1998) *Politicians and Firms in Seven Central and Eastern European Countries*, World Bank Technical Paper 1954, Washington DC.

Clark, D.P., J. Highfill, J. de Oliveira Campino, and S.S. Rehman (2011) 'FDI, Technology Spillovers, Growth, and Income Inequality: A Selective Survey', *Global Economy Journal* 11/2, pp. 1–42.

Coates, D. (ed.) (2005) *Varieties of Capitalism, Varieties of Approaches* (New York: Palgrave Macmillan).

Cohen, W. and D. Levinthal (1989) 'Innovation and Learning: The Two Faces of R&D', *The Economic Journal* 99/397, pp. 569–96.

Cohen, W. and D. Levinthal (1990) 'Absorptive capacity: a new perspective on learning and innovation', *Administrative Science Quarterly* 35/1, pp. 128–52.

Cohen, W., R. Nelson, and J. Walsh (2000) *Protecting their intellectual assets: Appropriability conditions and why US manufacturing firms patent (or not)*, NBER Working Paper 7552.

Contractor, F.J. (1990) 'Ownership Patterns of U.S. Joint Ventures Abroad and the Liberalization of Foreign Government Regulations in the 1980s: Evidence from the Benchmark Surveys', *Journal of International Business Studies* 21/1, pp. 55–73.

Cooke, P. et al. (1997) 'Regional innovation systems: Institutional and organisational dimensions', *Research Policy* 28, pp. 475–91.

Criscuolo, P., R. Narula, and B. Verspagen (2002) *The relative importance of home and host innovation systems in the internationalisation of MNE R&D: a patent citation analysis*, MERIT-Infonomics Research Memorandum 2002/26.

Czarnitzki, D. (2005) 'Extent and Evolution of the Productivity Deficiency in Eastern Germany', *Journal of Productivity Analysis* 24/2, pp. 209–29.

D'Cruz, J. R. (1986) 'Strategic management of subsidiaries', in H. Etemad and L. Seguin Dulude (eds) *Managing the Multinational Subsidiary* (London: Croom Helm) pp. 75–89.

Damijan, J.P. and M. Knell (2003a) *How important is trade and foreign ownership in closing the technology gap? Evidence from Estonia and Slovenia*, University of Ljubljana and Center for Technology, Innovation and Culture (TIK), Oslo, mimeo.

Damijan, J.P. and M. Knell (2003b) *Impact of privatization methods on the accessibility of local firms to international knowledge spillovers through trade and foreign investment: Evidence from Estonia and Slovenia*, University of Oslo, mimeo.

Damijan, J.P., Č. Kostevc, and M. Rojec (2011) 'Does a foreign subsidiary's network status affect its innovation activity? Evidence from post-socialist economies', *Economic and Business Review*, forthcoming.

Damijan, J.P., M. Knell, B. Majcen, and M. Rojec (2003a) 'The role of FDI, R&D accumulation and trade in transferring technology to transition countries: evidence from firm panel data for eight transition countries', *Economic Systems* 27, pp. 189–204.

Damijan, J.P., M. Rojec, B. Majcen, and M. Knell (2003b) *Technology Transfer through FDI in Top-10 Transition Countries: How Important are Direct Effects, Horizontal and Vertical Spillovers?*, William Davidson Working Paper 549, University of Michigan.

Damijan, J.P., M. Rojec, B. Majcen, and M. Knell (2008) *Impact of Firm Heterogeneity on Direct and Spillover Effects of FDI: Micro Evidence from Ten Transition Countries*, LICOS Discussion Paper218/2008

Das, S. (1987) 'Externalities and Technology Transfer Through Multinational Corporations', *Journal of International Economics* 22/1–2, pp. 171–182.

Disdier, A. and T. Mayer (2004) 'How different is Eastern Europe? Structure and determinants of location choices by French firms in Eastern and Western Europe' *Journal of Comparative Economics* 32, pp. 280–96.

Djankov, S. and B. Hoekman (1998) *Avenues of technology transfers: Foreign investment and productivity change in the Czech Republic*, CEPR Discussion Paper 1883.

Djankov, S. and B. Hoekman (2000) 'Foreign Investment and Productivity Growth in Czech Enterprises', *The World Bank Economic Review* 14/1, pp. 49–64.

Dries, L. and J.F.M. Swinnen (2004) 'Foreign Direct Investment, Vertical Integration, and Local Suppliers: Evidence from the Polish Dairy Sector', *World Development* 32/9, pp. 1525–44.

Driffield, N. (2006) 'On the Search for Spillovers from Foreign Direct Investment (FDI) with Spatial Dependency', *Regional Studies* 40 /1, pp. 107–19.

Driffield, N., M. Munday, and A. Roberts (2004) 'Inward investment, transaction linkages and productivity spillovers', *Papers in Regional Science* 83, pp. 699–722.

Dunning, J. and R. Narula (1995) 'The R&D activities of foreign firms in the United States', *International Studies of Management and Organization* 25/1–2, pp. 39–73.

Dunning, J. and S.M. Lundan (2008) *Multinational Enterprises and the Global Economy. 2nd ed.* (Cheltenham: Edward Elgar).

Dunning, J. H. (1988) 'The eclectic paradigm of international production: A restatement and some possible extensions', *Journal of International Business Studies* 19, pp. 1–31.

Dunning, J. H. (1993) *Multinational Enterprises and the Global Economy* (Addison-Wesley).

Dunning, J. H. and C. Wymbs (1999) 'The Geographical Sourcing of Technology-Based Assets by Multinational Enterprises', in D. Archibugi et al. (eds) *Innovation Policy in a Global Economy* (Cambridge: Cambridge University Press), pp. 185–224.

Dunning, J.H. (1977) 'Trade, Location of Economic Activity and MNE: A Search for an Eclectic Approach', in B. Ohlin, P.-O. Hesselborn, and P.M. Wijkman (eds) *The International Allocation of Economic Activity* (London: Macmillan) pp. 395–418.

Dunning, J.H. (1980) 'Toward an eclectic theory of international production: Some empirical tests', *Journal of International Business Studies* 11/1, pp. 9–31.

Dunning, J.H. (1981) *International Production and the Multinational Enterprise* (London: George Allen and Unwin).

Dunning, J.H. (1989) *Transnational Corporations and the Growth of Services: Some Conceptual and Theoretical Issues* (New York: United Nations).

Dunning, J.H. (2009) 'Location and the multinational enterprise: A neglected factor?' *Journal of International Business Studies*, 40, pp. 5–19.

Dunning, J.H. and M. Rojec (1993) *Foreign Privatization in Central & Eastern Europe*, CEEPN, Ljubljana, Slovenia.

Dyker, D.A. (2006) 'Contrasting Patterns in the Internationalisation of Supply Networks in the Motor Industries of Emerging Economies', *Post-Communist Economies* 18/2, pp. 189–204.

Dyker, D.A. and C. Stolberg (2003) *Productivity and capability in the transition economies – a historical and comparative perspective*, University of Sussex, mimeo.

Dyker, D.A. and N. von Tunzelmann (2002) *Network Alignment in Firms in Transition Countries: A Survey of Hungarian and Slovenian Companies*, University of Sussex, mimeo.

Dyker, D.A. et al. (eds) (2006), *Closing the EU East-West Productivity Gap* (London: Imperial College Press).

Eaton, J. and M. Gersovitz (1983) 'Country risk: Economic aspects', in R.J. Herring (ed.) *Managing International Risk* (Cambridge: Cambridge University Press).

Eaton, J. and S. Kortum (1996) 'Trade in Ideas: Patenting and Productivity in the OECD', *Journal of International Economics* 40, pp. 251–78.

Ebanks, K.D. (1989) 'Pirates of the Caribbean Revisited: The Examination of the Continuing Problem of Satellite Signal Piracy in the Caribbean and Latin America', *Law and Policy in International Business* 21, pp. 33–53.

Edler, J. and W. Polt (2008) 'International Industrial R&D – the policy challenges', Special issue of *Journal of Technology Transfer* 33/4.

Emmert, F., F. Kronthaler, and J. Stephan (2005) *Analysis of statements made in favour of and against the adoption of competition law in developing and transition economies*, IWH Sonderheft 1/2005, IWH, Halle.

Enderwick, P. (2005) 'Attracting "desirable" FDI: theory and evidence' *United Nations Conference on Trade and Development* 14/2, pp. 94–120.

Erdogan, A.I. (2011) 'Productivity Spillovers from Foreign Direct Investment: A Review of the Literature', *Middle Eastern Finance and Economics* 11, pp. 53–68.

Etzkowitz, H. (2008) *The Triple Helix – University-Industry-Government, Innovation in Action* (New York and London: Routledge).

Etzkowitz, H. and L. Leydesdorff (2000) The dynamics of innovation: from National Systems and "Mode 2" to a Triple Helix of university–industry–government relations, *Research Policy* 29, pp. 109–123.

Euronorm (2007) *Wachstumsdynamik und strukturelle Veränderungen der FuE-Potentiale im Wirtschaftssektor der Neuen Bundesländer* 9–11.

Evenett, S.J. (2005) *What is the Relationship between Competition Law and Policy and Economic Development?* University of Oxford, mimeo.

Evenett, S.J. and A. Voico (2002) *Picking Winners or Creating Them? Revisiting the Benefits of FDI in the Czech Republic*. World Bank series on international technology diffusion, mimeo.

Fagerberg, J. et al. (1994) *The dynamics of technology trade and growth* (Aldershot: Edward Elgar).

Faria, P. and W. Sofka (2008) *Formal and Strategic Appropriability Strategies of Multinational Firms – A Cross Country Comparison*, ZEW Discussion Paper 08–030, Mannheim.

Filatotchev, I., J. Stephan, and B. Jindra (2008) 'Ownership Structure, Strategic Controls and Exporting of Foreign-Invested Firms in Transition Economies', *Journal of International Business Studies* 39/7, pp. 1133–1148.

Forsgren, M. and T. Pedersen (1998) 'Centres of Excellence in Multinational Networks: The Case of Denmark', in J. Birkinshaw and N. Hood (eds) *Multinational Corporate Evolution and Subsidiary Development* (Basingstoke: Palgrave Macmillan) pp. 141–61.

Fosfuri, A., M. Motta, and T. Ronde (2001) 'Foreign Direct Investment and Spillovers Though Workers' Mobility', *Journal of International Economics* 53/1, pp. 205–22.

Fratocchi, L. and U. Holm (1998) 'Centres of excellence in the international firm', in J. Birkinshaw and N. Hood (eds) *Multinational Corporate Evolution and Subsidiary Development* (Basingstoke: Palgrave Macmillan) pp. 189–209.

Freeman, C. (1982) *The Economics of Industrial Innovation* (London: Frances Pinter).

Freeman, C. (1987) *Technology and Economic Performance: Lessons from Japan* (London: Pinter).

Fritsch, M., B. Görzig, O. Hennchen, and A. Stephan (2004 'Cost structure survey in Germany', *Schmollers Jahrbuch – Journal of Applied Social Sciences Studies* 124/4, pp. 557–66.

Frost, T. (2001) 'The geographic sources of foreign subsidiaries' innovation', *Strategic Management Journal* 22, pp. 101–23.

Fry, M.J. (1994) 'Foreign direct investment, financing and growth', in B. Fischer (ed.) Investment and Financing in Developing Countries (Baden-Baden: Nomos), pp. 181–214.

Fryges, H., S. Gottschalk, and K. Kohn (2010) 'The KfW/ZEW Start-up Panel: Design and Research Potential', *Schmollers Jahrbuch – Journal of Applied Social Sciences Studies* 130/1, pp. 117–131.

Fujita, M., P.R. Krugman, and A.J. Vanables (1999) *The spatial economy. Cities, regions, and international trade* (Cambridge: MIT Press).

Gauselmann, A. and B. Jindra (2010) 'Multinationale Unternehmen in den Neuen Ländern: Wandel in der Motivlage und differenzierte Wahrnehmung der Standortqualität', *Wirtschaft im Wandel* 6/2010, pp. 281–88.

Gauselmann, A. and M. Knell, and J. Stephan (2011a) 'What drives FDI into Central East Europe? Evidence from the IWH-FDI-Micro Database', *Post-Communist Economies* 23/3, pp. 343–58.

Gauselmann, A., P. Marek, and J. Angenendt (2011b) *The role of Labor markets in multinational firms' regional location choice in transition economies*, IWH-Discussion Paper, forthcoming.

Gerschenkron, A. (1962) *Economic Backwardness in Historical Perspective: a Book of Essays* (Cambridge, Mass.: Harvard University Press).

Ghizdeanu, I. and V. Tudorescu (2007) 'Unit Labor Cost in Romania', *Romanian Journal of Economic Forecasting* 1/2007, pp. 57–64.

Ghoshal, S. and H. Nohria (1989) 'Internal differentiation within multinational corporations', *Strategic Management Journal* 10/4, pp. 323–37.

Gibbons, M., et al. (1994) 'Evolution of Knowledge Production', *The New Production of Knowledge* (London: Sage) pp. 17–45.

Ginarte, J.C. and W.G. Park (1997) 'Determinants of patent rights: A cross-national study', *Research Policy* 26, pp. 283–301.

Girma, S. (2005a) 'Absorptive Capacity and Productivity Spillovers from FDI: A Threshold Regression Analysis', *Oxford Bulletin of Economics and Statistics* 67/3, pp. 281–306.

Girma, S. (2005b) 'Technology Transfer from Acquisition FDI and the Absorptive Capacity of Domestic Firms: An Empirical Investigation', *Open Economics Review* 14/1, pp. 165–78.

Girma, S. and H. Görg (2005) *Foreign Direct Investment, Spillovers and Absorptive Capacity: Evidence from Quantile Regressions*, Kiel Institute for World Economics Working Paper 1248.

Giroud, A. (2007) 'MNEs Vertical Linkages: The Experience of Vietnam after Malaysia', *International Business Review* 16/2, pp. 159–76.

Glass, A.J. (2003) *Intellectual Property Policy and Spillovers from Multinationals*, Department of Economics, Texas A&M University, mimeo.

Glass, A.J. and K. Saggi (2002a) 'Multinational Firms and Technology Transfer', *Scandinavian Journal of Economics* 104/4, pp. 495–503.

Glass, A.J. and K. Saggi (2002b) 'Intellectual property rights and foreign direct investment', *Journal of International Economics* 56, pp. 387–410.

Globerman, S. (1979) 'Foreign Direct Investment and Spillover Efficiency Benefits in Canadian Manufacturing Industries', *Canadian Journal of Economics* 12/1, pp. 42–56.

Görg, H. and D. Greenaway (2003) *Much ado about nothing? Do domestic firms really benefit from FDI?*, IZA Discussion Paper 944, Bonn.

Görg, H. and D. Greenaway (2004), 'Much ado about nothing? Do domestic firms really benefit from foreign direct investment?', *World Bank Research Observer* 19, pp. 171–97.

Görg, H. and E. Strobl (2001) 'Multinational Companies and Productivity Spillovers: A Meta-analysis', *Economic Journal* 111, pp. F723–F739.

Görg, H., A. Hijzen, and B. Muraközy (2009) *The role of production technology for productivity spillovers from multinationals: Firm-level evidence for Hungary*, Kiel Working Paper 1482, February 2009.

Granovetter, M. (1985) 'Economic action and social structure: The problem of embeddedness', *American Journal of Sociology* 91, pp. 10–481.

Granovetter, M. (1992) 'Problems of explanation in economic sociology', in N. Nohria and R. Eccles (eds) *Networks and organizations: Structure, form and action* (Boston: Harvard Business School Press).

Greene, W. H. (2003) *Econometric Analysis* (New Jersey, Upper Saddle River: Pearson Education International).

Günther, J. (2005) 'The Absence of Technological Spillovers from Foreign Direct Investment in Transition Economies', in P.J.J. Welfens and A. Zwiatek-Kubiak (eds) *Structural Change and Exchange Rate Dynamics – The Economics of the EU Eastern Enlargement* (Berlin: Springer), pp. 149–66.

Günther, J. and O. Gebhardt (2005) 'Eastern Germany in the Process of Catching-up – The Role of Foreign and West German Investors for Technological Renewal', *Eastern European Economics* 43/5, pp. 78–102.

Günther, J., A. Gauselmann, P. Marek, J. Stephan, and B. Jindra (2011a) 'An Introduction to the IWH FDI Micro Database', *Schmollers Jahrbuch – Journal of Applied Social Science Studies* 131/3, pp. 529–46.

Günther, J., B. Jindra, and J. Stephan (2008a) *Foreign Subsidiaries in the East German Innovation System – Evidence from manufacturing industries*, IWH Discussion Paper 4/2008, Halle.

Günther, J., J. Stephan, and B. Jindra (2008b) 'Foreign Subsidiaries in the East German Innovation System – Evidence from Manufacturing Industries', *Applied Economics Quarterly Supplement* 59, pp. 137–65.

Günther, J., J. Stephan, and B. Jindra (2009) 'Does Local Technology matter for Foreign Investors in Central and Eastern Europe? – Evidence from the IWH FDI Micro Database', Special Issue of the *Journal of East West Business* on Market Entry and Operational Strategies of MNEs in Transition and Emerging Economies 15/3&4, pp. 210–47.

Günther, J., J. Stephan, and B. Jindra (2011b) 'FDI and the National Innovation System – Evidence from Central and Eastern Europe', in D. Dyker (ed.) *Network Dynamics in Emerging Regions of Europe* (Imperial College Press), pp. 303–32.

Gupta, A. K. and V. Govindarajan (1994) 'Organizing for knowledge within MNCs', *International Business Review* 3/4, pp. 443–57.

Håkansson, H. (1989) *Corporate technological behaviour. Cooperation and networks.* (London: Routledge).

Hall, P. A. and D. Soskice (eds) (2001), *Varieties of Capitalism. The Institutional Foundations of Comparative Advantage* (Oxford: Oxford University Press).

Halpern, L. and B. Muraközy (2009), 'Innovation, productivity and exports: the case of Hungary', in wiiw MICRO-DYN Deliverable 47, EU 6th Framework Programme, mimeo.

Hamar, J. and J. Stephan (2005) 'FDI, Productivity and Economic Restructuring in Central and Eastern Europe', in J. Stephan (ed.), *Foreign Direct Investment and Technology Transfer in Transition Countries: Theory – Method of Research – Empirical Evidence* (Basingstoke: Palgrave Macmillan and New York: St Martin's Press), pp. 77–95.

Hamar, J. and J. Stephan (2005) 'The Potentials for Technology Transfer via Foreign Direct Investment in Central East Europe – Results of a Field Study', *East-West Journal of Economics and Business* VIII/1&2, pp. 19–41.

Hamar, J. and J. Stephan (2006) 'Control of FDI subsidiaries and reciprocal technology transfer', in H. Brezinski and B. Leick (eds), *Kooperationsperspektiven deutscher Unternehmen in östlichen Grenzregionen* (Freiberg: TU Bergakademie Freiberg), pp. 69–86.

Harding, T. and B.S. Javorcik (2011) 'Roll out the red carpet and they will come: investment promotion and FDI inflows', *The Economic Journal* 121/557, pp. 1445–76.

Harhoff, D. and G. Licht (1993) *Das Mannheimer Innovations Panel*, ZEW Discussion Paper 93–21, Mannheim.

Hartwig, K.-H. (1985) 'Konzeption der Genossenschaften in Osteuropa', in E. Boettcher (ed.), *Die Genossenschaften im Wettbewerb der Ideen* (Tübingen: Mohr und Siebeck), pp. 213–31.

Haskel, J.E., S.C. Pereira, and M.J. Slaughter (2002) *Does Inward Foreign Direct Investment Boost the Productivity of Domestic Firms?*, NBER Working Paper 8724 (Cambridge, MA: National Bureau of Economic Research, Inc.).

Hatzichronoglou, T. (1997) *Revisions of the high-technology sector and product classification*, OECD STI Working Paper 2, Paris.

Hayek, F.A. von (1948) 'The Meaning of Competition', in F.A. von Hayek, *Individualism and Economic Order* (London: Routledge), pp. 92–106.

Hayek, F.A. von (1968) 'Competition as a Discovery Procedure', in F.A. von Hayek (1978), *New Studies in Philosophy, Politics and Economics* (Chicago: University of Chicago Press), pp. 179–90.

Hedlund, G. (1986) 'The hypermodern MNC: A heterarchy?' *Human Resources Management* 25, pp. 9–36.

Helpman, E. (2006) 'Trade, FDI, and the Organization of Firms', *Journal of Economic Literature* 44/3, pp. 589–630.

Helpman, E., M. J. Melitz, and S.R. Yeaple (2004) 'Export versus FDI with heterogeneous firms', *The American Economic Review* 94, pp. 300–16.

Herr, H. and A. Westphal (1991) 'Die Inkohärenzen der Planwirtschaft und der Transformationsprozeß zur Geldwirtschaft', in J. Backhaus (eds), *Systemwandel und Reform in östlichen Wirtschaften* (Marburg: Metropolis), pp. 139–68.

Hirsch, S. (1965) 'The United States electronics industry in international trade', *National Institute Economic Review* 24, pp. 92–7.

Hirsch, S. (1967) *Location of Industry and International Competitiveness* (Oxford: Clarendon Press).

Hoekman, B. and B. Smarzynska Javorcik (2006) 'Lessons from empirical research on international technology diffusion through trade and foreign direct investment', in B. Hoekman and B. Javorcik Smarzynska (eds) *Global Integration and Technology Transfer* (Basingstoke: Palgrave MacMillan, New York: The World Bank).

Holm, U., A. Malmberg, and Ö. Sölvell (2002) *MNC Impact on Local Clusters*, IIB Institute of International Business Working Paper 02/11, IIB: Stockholm.

Hölscher, J. and 1 (1997) 'Export-Oriented Development and Income Creation in Hungary', in J.G. Backhaus and G. Krause (eds), *The Political Economy of Transformation: Country Studies* (Marburg: Metropolis), pp. 47–71.

Hölscher, J. and J. Stephan (2004) 'Competition Policy in Central East Europe in Light of EU Accession', *Journal of Common Market Studies* 42/2, pp. 321–45.

Hölscher, J. and J. Stephan (2008) 'Competition and Antitrust Policy in the Enlarged European Union – A Level Playing Field?', *Journal of Common Market Studies* 47/4, pp. 863–89.

Hoskisson, R.E., L. Eden, C. Ming Lao, and M. Wright (2000) 'Strategy in Emerging Economies', *Academy of Management Journal* 43/3, pp. 249–67.

Howells, J. and M. Wood (1993) *The Globalisation of Production and* Technology (London: Belhaven Press).

Hu, Y.-S., (1992) 'Global or stateless corporations are national firms with international operations', *California Management Review* Winter, pp. 107–26.

Hunya, G. (1997) 'Large privatisation, restructuring and foreign direct investment', in S. Zecchini (ed.) *Lessons from the Economic Transition. Central and Eastern Europe in the 1990s* (Dordrecht: Kluwer Academic Publishers) pp. 275–300.

Hunya, G. (2010) *wiiw FDI database on FDI in Central, East and Southeast Europe: FDI in the CEECs Hit Hard by the Global Crisis* (Vienna: wiiw).

Hymer, S. H. (1976) *The International Operations of National Firms: A Study of Direct Investment* (Cambridge, MA: MIT Press).

Ietto-Gilles, G. (2005) *Transnational Corporations and International Production – Concepts, Theories and Effects* (Cheltenham: Edward Elgar).

Imbriani, C. and F. Reganti (1999) *Productivity Spillovers and Regional Differences: Evidence from the Italian Manufacturing Sector,* Centro del Economica de Lavoro e de Politica Economica, University of Salerno, Discussion Paper 48.

IWH (2007), *IWH FDI Micro Database – Methodological Note Survey in Transition countries,* online publication, Halle Institute of Economic Research, Halle.

IWH (2009), *IWH FDI Micro Database – Methodological Note Survey in Transition countries,* online publication, Halle Institute of Economic Research, Halle.

Jaffe, A. B., M. Trajtenberg, and R. Henderson (1993) 'Geographical localisation of knowledge spillovers as evidence by patent citations', *Quarterly Journal of Economics* 108, pp. 577–98.

Jaffe, A.B. and M. Trajtenberg (1998) *International Knowledge Flows: Evidence from Patent Citations,* NBER Working Papers 6507, National Bureau of Economic Research, Inc.

Jaffe, A.B., M.S. Fogarty, and B.A. Banks (1998) 'Evidence from Patents and Patent Citations on the Impact of NASA and Other Federal Labs on Commercial Innovation', *The Journal of Industrial Economics* 46, pp. 183–205.

Jandhyala, S. (2008) *De facto Property Rights Protection and MNC Location Choices,* The Wharton School, University of Pennsylvania, mimeo.

Janz, N. (ed.) (2003) *Innovationsforschung heute* (Nomos: Baden-Baden).

Jarillo, J.-C. and J. I. Martinez (1990) 'Different roles for subsidiaries: The case of multinational corporations', *Strategic Management Journal* 11/7, pp. 501–12.

Jensen, C. (2002) *Spillovers in the Polish Food Industry. Exploring the role of local externalities and global networks.* Working paper presented at the 3rd CEES Workshop on Transition and Enterprise Restructuring, Copenhagen Business School, 15–17 August, 2002.

Jindra, B. (2006) 'Part I: Theory and Review of the latest Research on the Effects of FDI into Central East Europe', in Stephan (ed.) *Technology Transfer via Foreign Direct Investment in Central and Eastern Europe – Theory, Method of Research and Empirical Evidence,* Studies in Economic Transition (Houndsmills: Palgrave Macmillan) pp. 3–75.

Jindra, B. (2011) *Internationalisation Theory and Technological Accumulation– An investigation of multinational affiliates in East Germany.* Studies in Economic Transition edited by J. Hölscher and H. Tomann (Basingstoke: Palgrave Macmillan).

Jindra, B. and M. Rojec (2012) 'Foreign Direct Investment and Knowledge Spillovers in Transition Economies – Is there a need for more policy coordination?' Forthcoming in a special issue of *Journal of the Knowledge Economy,* 2012.

Jungmittag, A. (2005) 'Innovations, Technological Specialisation and Economic Convergence in the EU', in P.J.J. Welfens and A. Zwiatek-Kubiak (eds) *Structural Change and Exchange Rate Dynamics – The Economics of the EU Eastern Enlargement* (Berlin: Springer), pp. 171–199.

Kalotay, K. (2010) 'Patterns of inward FDI in economies in transition', *Eastern Journal of European Studies* 1/2, pp. 55–76.

Karhunen, P. J. Löfgren, and R. Kosonen (2008) 'Revisiting the relationship between ownership and control in international business operations: lessons from transition economies', *Journal of International Management* 14, pp. 78–88.

Keller, W. (1996) 'Absorptive capacity: On the creation and acquisition of technology in development', *Journal of Development Economics* 49/1, pp. 199–227.

Keller, W. and S.R. Yeaple (2003) *Multinational Enterprises, International Trade, and Productivity Growth: Firm-Level Evidence from the United States*, IMF Working Paper 03/248 (Washington D.C.: International Monetary Fund).

Keynes, J.M. (1936) *The General Theory of Employment, Interest and Money* (London: Macmillan).

Kinoshita, Y. (2000) *R&D and technology spillovers via FDI: Innovation and absorptive capacity*, Working Paper 349, The William Davidson Institute, University of Michigan Business School.

Kleinknecht, A. (1989) 'Firm size and innovation, Observations in Dutch Manufacturing Industries', *Small Business Economics* 1, pp. 215–22.

Klevorick, A.K., R.C. Levin, R.R. Nelson, and S.G. Winter (1995) 'On the sources and significance of inter-industry differences in technological opportunities', Research Policy 24, pp. 185–205.

Kneller, R. (2002) *Frontier Technology, Absorptive Capacity And Distance*, Research Paper Series: Globalisation, Productivity and Technology, 2002 (24), Leverhulme Centre.

Kogut, B. and H. Singh (1988) 'The Effect of National Culture on the Choice of Entry Mode', *Journal of International Business Studies* 19/3, pp. 411–32.

Kogut, B. and U. Zander (1992) 'Knowledge of the firm, combinative capabilities, and the replication of technology', *Organization Science* 3/3, pp. 383–97.

Kogut, B. and U. Zander (1993) 'Knowledge of the firm and the evolutionary theory of the multinational corporation', *Journal of International Business Studies* 24, pp. 625–45.

Kogut, B. and U. Zander (1996) 'What do firms do? Coordination, identity and learning', *Organization Science* 7, pp. 502–518.

Kokko A., R. Tansini, and M. Zejan (1996) 'Local Technological Capability and Productivity Spillovers from FDI in the Uruguayan Manufacturing Sector', *Journal of Development Studies* 32/4, pp. 602–11.

Kokko, A. (1994) 'Technology, Market Characteristics and Spillovers', *Journal of Development Economics* 43, pp. 279–93.

Kokko, A. and V. Kravtsova (2008) 'Innovative capability in MNC subsidiaries: Evidence from four European transition economies', *Post-Communist Economies* 20/1, pp. 57–5.

Kolasa, M. (2007) *How does FDI inflow affect productivity of domestic firms? The role of horizontal and vertical spillovers, absorptive capacity and competition*, National Bank of Poland Working Paper 42, Warsaw.

Konings, J. (2001) 'The Effects of Foreign Direct Investment on Domestic Firms. Evidence from Firm-level Panel Data in Emerging Economies', *Economics of Transition* 9/3, pp. 619–33.

Koschatzky, K. et al. (2006) *Innovationsbedingungen und Innovationspotenziale in Ostdeutschland– Exemplarische Analyse von drei Grenzregionen* (Stuttgart: Fraunhofer IRB Verlag).

Krugman, P. (1996) 'What's new about the new economic geography?', *Oxford Review of Economic Policy* 14, pp. 7–17.

Kuemmerle, W. (1997) 'Building effective R&D capabilities abroad', *Harvard Business Review* 3/4, pp. 61–70.

Kvinge, T. (2004) *Knowledge diffusion through FDI – established wisdom or wishful thinking?* Centre for technology, innovation and culture, University of Oslo, Working paper 31.

Lall, S. (1980) 'Vertical Inter-Firm Linkages in LDCs: An Empirical Study', *Bulletin of Economics and Statistics* 42/3, pp. 203–26.

Lane, P.J. and M. Lubatkin (1998) 'Relative absorptive capacity and interorganizational learning', *Strategic Management Journal* 19/5, pp. 461–77.

Lankes, H., and A. Venables (1996) 'Foreign direct investment in economic transition: The changing pattern of investments', *Economics of Transition* 4/2, pp. 331–347.

Lavigne, M. (1998) 'Conditions for accession to the EU', *Comparative Economic Studies* 40, pp. 38–57.

Lavigne, M. (1999) *The Economics of Transition: From Socialist Economy to Market Economy* (New York, N.Y.: St. Martin's Press).

Lee, J.Y. and E. Mansfield (1996) 'Intellectual Property Protection and U.S. Foreign Direct Investment', *The Review of Economics and Statistics* 78, pp. 181–86.

Léger, A. (2007) *The Role(s) of Intellectual Property Rights for Innovation: A Review of the Empirical Evidence and Implications for Developing Countries*, Deutsches Institut für Wirtschaftsforschung Berlin Discussion Paper 707, DIW, Berlin.

Legler, H. et al. (2007) 'Die Bedeutung von Aufhol-Ländern im globalen Technologiewettbewerb', *Studien zum deutschen Innovationssystem* 2007(21).

Leiponen, A. and C.E. Helfat (2004) *Innovation objectives, knowledge sources, and the benefits of breadth* (Cornell: Cornell University).

Levin, R., A. Klevorick, R.R. Nelson, and S.G. Winter (1987) 'Appropriating the returns from industrial R&D' *Brookings Papers on Economic Activity*, pp. 783–820.

Levitt, B. and J.G. March (1987) 'Organizational learning' *Annual Review of Sociology* 14, pp. 319–40.

Leydesdorff, L. (2004) *Measuring the knowledge base of an economy in terms of triple-helix relations among 'technology, organization, and territory*, Erasmus Research Institute of Management Rotterdam: ERIM.

Lipsey, R.E. and F. Sjöholm (2005) 'The Impact of FDI on Host Countries: Why Such Different Answers', in T.H. Moran, E.M. Graham, and M. Blomström (eds), *Does Foreign Direct Investment promote Development?* (Washington, DC: Institute for International Economics and Center for Global Development), pp. 23–44.

Lipton, D. and J.D. Sachs (1990) 'Creating a Market Economy in Eastern Europe: The Case of Poland', in *Brookings Papers on Economic Activity* 1, pp. 75–147.

Love, J.H. and M.A. Mansury (2009) 'Exporting and productivity in business services: Evidence from the United States', *International Business Review* 18/6, pp. 630–42.

Lundvall, B.-Å. (1988) 'Innovation as an interactive process: from user–producer interaction to the national system of innovation', in G. Dosi, C. Freeman, R. Press Nelson, G. Silverberg, and L. Soete (eds) *Technical change and economic theory* (London: Pinter).

Lundvall, B.-Å. (ed.) (1992) *National Innovation Systems: Towards a Theory of Innovation and Interactive Learning* (London: Pinter).

Lyles, M. and J. Salk (1996) 'Knowledge acquisition from foreign partners in international joint ventures', *Journal of International Business Studies* 27, pp. 877–904.

Majcen, B., S. Radošević, and M. Rojec (2003c) *FDI subsidiaries and industrial integration of Central Europe: Conceptual and empirical results*, IWH Discussion Paper 177, Halle Institute for Economic Research, Halle.

Majocchi, A. and R. Strange (2007) 'The FDI location decision: does liberalization matter?' *Transnational Corporations* 16/2, pp. 1–40.

Major, I. (2003) 'Privatisation in Hungary and Its Aftermath', in D. Parker and D. Saal (eds), *International Handbook on Privatisation* (Cheltenham: Edward Elgar), pp. 427–53.

Manea, J. and R. Pearce (2004) 'Industrial restructuring in economies in transition and TNCs' investment motivations', *Transnational Corporations* 13/2, pp. 7–27.

Manea, J. and R. Pearce (2006) 'MNE strategies in CEE: Key elements of subsidiary behaviour', *International Management Review* 46/2, pp. 435–55.

Mansfield, E. (1986) 'Patents and Innovations: An Empirical Study', *Management Science* 32/2, pp. 173–81.

Mansfield, E. (1991) 'Academic research and industrial innovation', *Research Policy* 20, pp. 1–12.

Mansfield, E. (1993) 'Unauthorized Use of Intellectual Property: Effects on Investment, Technology Transfer, and Innovation', in M.B. Wallerstein, M.E. Mogee, and R.A. Schoen (eds) *Global dimensions of intellectual property rights in science and technology, Office of International Affairs, National Research Council* (Washington, D.C : National Academic Press).

Mansfield, E. (1995) *Intellectual Property Protection, Direct Investment, and Technology Transfer: Germany, Japan, and the United States*, International Finance Corporation Discussion Paper 27, The World Bank, Washington, D.C.

Mansfield, E. and M. Romeo (1980) 'Technology Transfer to Overseas Subsidiaries by U.S. Based Firms', *Quarterly Journal of Economics* 95, pp. 737–50.

March, J.G. (1991) 'Exploration and exploitation in organizational learning', *Organization Science* 2/1, pp. 71–87.

Marcotte, C. and J. Niosi (2000) 'Technology Transfer to China – The Issues of Knowledge and Learning', *Journal of Technology Transfer* 25, pp. 43–57.

Marin, A. and M. Bell (2006) 'Technology Spillovers from foreign direct investment: an exploration of the active role of MNE subsidiaries in the case of Argentina in the 1990s', *Journal of Development Studies* 42/4, pp. 678–97.

Markusen, J.R. and A.J. Venables (1999) 'Foreign Direct Investment as a Catalyst for Industrial Development', *European Economic Review* 43/2, pp. 335–56.

Marshall, A. (1962) *Principles of Economics. An introductory volume. 8th edition (first edition in 1890)* (London: MacMillan).

Maskus, K.E. (1998) 'The Role of Intellectual Property Rights in Encouraging Foreign Direct Investment and Technology Transfer', *Duke Journal of Comparative and International Law* 9, pp. 109–61.

Maskus, K.E. (2000) *Intellectual Property Rights and Foreign Direct Investment*, Policy Discussion Paper 22, Centre for International Economic Studies, University of Adelaide.

Maskus, K.E., K. Saggi, and T. Puttitanun (2004) *Patent Rights and International Technology Transfer though Direct Foreign Investment and Licensing.* Paper prepared for the conference 'International Public Goods and the Transfer of Technology after TRIPS', Duke University Law School April 4–6, 2003.

Maurseth, P.B. and B. Verspagen (2002) 'Knowledge Spillovers in Europe: A Patent Citations Analysis', *Scandinavian Journal of Economics* 104/4, pp. 531–45.

McCalman, P. (2001) 'Reaping What You Sow: An Empirical Analysis of International Patent Harmonization', *Journal of International Economics* 55/1, pp. 161–86.

McGowan, S., N. Radosevic, N. von Tunzelmann (2004) *The Emerging Structure of the wider Europe* (London: Routledge).

Melitz, M.J. (2003) 'The Impact of Trade on Intra-Industry Reallocations and Aggregate Industry Productivity', *Econometrica* 71/6, pp. 1695–725.

Meske, W. (2004) *From System Transformation to European Integration. Science and technology in Central and Eastern Europe at the beginning of the 21st century* (Münster: Lit Verlag).

Metcalfe, S. (1995) 'The Economic Foundations of Technology Policy: Equilibrium and Evolutionary Perspectives', in P. Stoneman (ed.), *Handbook of the Economics of Innovation and Technological Change* (Oxford (UK) and Cambridge (US): Blackwell Publishers).

Meyer, K. and E. Sinani (2009) 'When and Where Does Foreign Direct Investment Generate Positive Spillovers? A Meta-Analysis', *Journal of International Business Studies* 40/7, pp. 1075–1094.

Meyer, K.E. (1995) 'Direct foreign investment in Eastern Europe. The role of labor costs', *Comparative Economics Studies* 37, pp. 69–88.

Meyer, K.E. (1998) *Direct Investment in Economies in Transition* (Edward Elgar: Cheltenham, UK and Northampton, MA, USA).

Meyer, K.E. (2001) 'Institutions, transaction costs, and entry mode choice in Eastern Europe', *Journal of International Business Studies* 32/2, pp. 357–67.

Meyer, K.E. (2003) *FDI Spillovers in Emerging Markets: A Literature Review and New Perspectives*, Copenhagen Business School, mimeo.

Meyer, K.E. and C. Jensen (2004) *Foreign Direct Investment and Government Policy in Central and Eastern Europe*, mimeo.

Meyer, K.E. and M.W. Peng (2005) 'Probing Theoretically into Central and Eastern Europe: Transactions, Resources, and Institutions', *Journal of International Business Studies* 36/6, pp. 600–21.

Mickiewicz, T. and M. Baltowski (2003) 'All Roads Lead to Outside Ownership: Polish Piecemeal Privatisation', in D. Parker and D. Saal (eds) *International Handbook on Privatisation* (Cheltenham: Edward Elgar), pp. 402–26.

Miller, R. (1994) 'Global R&D networks and large-scale innovations: the case of the automobile industry', *Research Policy* 23, pp. 27–46.

Mishra, C.S. and R.K. Zachary (2011) 'Revisiting, Reexamining and Reinterpreting Schumpeter's Original Theory of Entrepreneurship', *Entrepreneurship Research Journal* 1/1, pp. 1–6.

Moran, T.H. and C.F. Bergsten (1998) *Foreign direct investment and development: the new policy agenda for developing countries and economics of transition* (Washington DC: Institute for International Economics).

Moran, T.H., E.M. Graham, and M. Blomström (eds) (2005), *Does Foreign Direct Investment promote Development?* (Washington, DC: Institute for International Economics and Center for Global Development).

Mowery, D. and N. Rosenberg (1979) 'The influence of market demand upon innovation: a critical review of some recent empirical studies', *Research Policy* 8, pp. 102–53.

Mudambi, R. (1999) 'MNE internal capital markets and subsidiary strategic independence', *International Business Review* 8/2, pp. 197–211.

Mudambi, R. and S.M. Mudambi (2002) 'Diversification and market entry choices in the context of foreign direct investment', *International Business Review* 11, pp. 35–55.

Mutinelli, M. and L. Piscitello (1997) 'Differences in the strategic orientation of Italian MNEs in Central and Eastern Europe: The influence of firm-specific factors', *International Business Review* 6/2, pp. 185–205.

Myant, M. and J. Drahokoupil (2012), 'International integration, varieties of capitalism, and resilience to crisis in transition economies', forthcoming in *Europe-Asia Studies*.

Nahapiet, J. and S. Ghoshal (1998) 'Social capital, intellectual capital, and the organizational advantage', *Academy of Management Review* 23/2, pp. 242–66.

Narula, R. (2002) 'Innovation systems and 'inertia' in R&D location: Norwegian firms and the role of systemic lock-in', *Research Policy* 31, pp. 795–816.

Nelson, R.R. (2008) 'Economic Development from the Perspective of Evolutionary Economic Theory', *Oxford Development Studies* 36/1, pp. 9–21.

Nelson, R.R. (ed.) (1993) *National Innovation Systems. A Comparative Analysis* (New York/Oxford: Oxford University Press).

Nelson, R.R. and S.G. Winter (1982) *An evolutionary theory of economic change* (Cambridge, MA: Harvard University Press).

Nicholson, M.W. (2007) 'The Impact of Industry Characteristics and IPR Policy on Foreign Direct Investment', *Review of World Economics* 143/1, pp. 27–54.

Nonaka, I. (1994) 'Dynamic Theory of Organizational Knowledge Creation', *Organization Science* 5/1, pp. 14–37.

Noorderhaven, N.A. and W. Harzing (2003) 'The "country-of-origin effect" in multinational corporations: Sources, mechanisms and moderating conditions', *Management International Review* 43, pp. 47–66.

Nunnenkamp, P. and J. Spatz (2003) *Intellectual Property Rights and Foreign Direct Investment: The Role of Industry and Host-Country Characteristics*, Kiel Institute for World Economics Working Paper 1167, Kiel.

OECD (1999) *Science, Technology and Industry Scoreboard 1999. Benchmarking Knowledge-Based Economies*, Paris.

OECD (2002) *Frascati Manual – Proposed standard practise for survey on research and experimental development* (Paris: OECD).

OECD (2005) *Oslo Manual-Guidelines for collecting and interpreting innovation data, 3rd edition* (Paris: OECD and Eurostat).

Ohmae, K. (1990) *The Borderless World: Power and Strategy in the Interlinked Economy* (London: Collins).

Oxley, J.E. (1999) 'Institutional environment and the mechanisms of governance: the impact of intellectual property protection on the structure of inter-firm alliances', *Journal of Economic Behavior & Organization* 38/3, pp. 283–309.

Ozawa, T. (1979) *Multinationalism, Japanese Style* (Princeton: Princeton University Press).

Pack, H. and K. Saggi (2001) 'Vertical Technology Transfer and Transfer via International Outsourcing', *Journal of Development Economics* 65/2, pp. 389–415.

Park, W. and D.C. Lippoldt (2003) *The impact of trade-related intellectual property rights on trade and foreign direct investment in developing countries* (Paris: OECD).

Park, W. and D.C. Lippoldt (2005) *Technology Transfer and the Economic Implications of the Strengthening of Intellectual Property Rights in Developing Countries*, OECD Trade Policy Working Paper 62, Paris.

Pasternack, P. (2007) Forschungslandkarte Ostdeutschland, Sonderband "die hochschule" 2007 (Wittenberg: Institut für Hochschulforschung).

Patel, P. and K. Pavitt (1991) 'Large firms in the production of the world's technology: An important case of "Non-Globalisation" ', *Journal of International Business Studies*, 22/1, pp. 1–21.

Patel, P. and K. Pavitt (1994) *The Nature and Economic Importance of National Innovation Systems*, STI Review, No. 14 (Paris: OECD).

Patel, P. and K. Pavitt (1999) 'Global Corporations & National Systems Of Innovation: Who Dominates Whom?', in D. Archibugi, J. Howells, J. Michie (eds) *Innovation Policy in a Global Economy* (Cambridge: Cambridge University Press).

Patel, P. and K. Pavitt (2000) 'National systems of innovation under strain: the Internationalisation of Corporate R&D', in R. Barrell, G. Mason, and M. O'Mahoney (eds), *Productivity, Innovation and Economic Performance* (Cambridge: Cambridge University Press).

Patel, P. and M. Vega (1999) 'Patterns of internationalisation and corporate technology: location versus home country advantages', *Research Policy* 28, pp. 145–55.

Paul, D.L. and R.B. Wooster (2008) 'Strategic investments by US firms in transition economies', *Journal of International Business Studies* 39, pp. 249–66.

Pauly, L.W. and S. Reich (1997) 'National structures and multinational corporate behavior: enduring differences in the age of globalization', *International Organization* 51, pp. 1–30.

Pavitt, K. (1998) 'The Social Shaping of the National Science Base', *Research Policy* 27, pp. 793–805.

Pavitt, K. (2005) 'Innovation Processes', in J. Fagerberg, D.C. Mowery, and R.R. Nelson (eds) *The Oxford Handbook of Innovation* (Oxford: Oxford University Press), pp. 86–114.

Pearce, R.D and M. Papanastassiou (1999) 'Overseas R&D and the strategic evolution of MNEs: evidence from laboratories in the UK', *Research Policy* 28, pp. 23–41.

Pearce, R.D and S. Singh (1992) *Globalizing Research and development* (London: Macmillan).

Pellegrin, J. (1999) *German Production Networks in Central/Eastern Europe between Dependency and Globalisation* (Berlin: Wissenschaftscentrum Berlin für Sozialforschung).

Peng, M.W. (2003) 'Institutional transitions and strategic choices', *Academy of Management Review* 28/2, pp. 275–96.

Pitelis, C.N. (1998), 'Transaction Costs and the Historical Evolution of the Capitalist Firm', *Journal of Economic Issues* XXXII/4, pp. 999–1017.

Polanyi, K. (1957) *The great transformation* (Boston: Beacon Press).

Porter, M. E. (1990). *The Competitive Advantage of Nations* (New York: Free Press).

Powell, W., K. Koput, and L. Smith-Doerr (1996): 'Interorganizational collaboration and the locus of innovation: networks of learning in biotechnology', *Administrative Science Quarterly* 41/1, pp. 116–145.

Pusterla, F. and L. Resmini (2007) 'Where do foreign firms locate in transition countries? An empirical investigation', *The Annals of Regional Science* 41/4, pp. 835–56.

Reddy, N.M. and L. Zhao (1990) 'International technology transfer: a review', *Research Policy* 19, pp. 285–307.

Resmini, L. (2000) 'The determinants of foreign direct investment in the CEECs. New Evidence from sectoral Patterns', *Economics of Transition* 8/3, pp. 665–89.

Rhee, Y.W. (1990) 'The Catalyst Model of Development: Lessons from Bangladesh's Success with Garment Exports', *World Development* 182/2, pp. 333–46.

Riese, H. (1991) 'Geld und Systemfrage', in J. Backhaus (ed.), *Systemwandel und Reform in östlichen Wirtschaften* (Marburg: Metropolis), pp. 125–38.

Riese, H. (1992) 'Das Scheitern des Sozialismus und der Transformationsprozeß', in A. Schikora, A. Fiedler, and E. Hein (eds), *Politische Ökonomie im Wandel. Festschrift für K.P. Kisker* (Marburg: Metropolis), pp. 23–36.

Rockett, K. (1990) 'The Quality of Licensed Technology', *International Journal of Industrial Economics* 8, pp. 559–74.

Rodriguez-Clare, A. (1996) 'Multinationals, Linkages, and Economic Development' *American Economic Review* 86/4, pp. 852–73.

Rodrik, D. (2006) 'Goodbye Washington Consensus, Hello Washington Confusion?' *Journal of Economic Literature* XLIV, December, pp. 969–83.

Rojec, M. and M. Svetličič (1993) 'Foreign direct investment in Slovenia', *Transnational Corporations* 2/1, pp. 135–51.

Roth, D. (2006) 'Wissenschaftseinrichtungen als Standortfaktor', *IWH Sonderheft* 2006/04.

Roth, K. and J. Morrison (1992) 'Implementing global strategy: Characteristics of global subsidiary mandates', *Journal of International Business Studies* 23/4, pp. 715–36.

Rugman, A. M. and J.P. Doh (2008) *Multinationals and Development* (New Haven & London: Yale University Press).

Rugraff, E. (2008) 'Are the FDI policies of the Central European Countries efficient?', *Post-Communist Economies* 20/3, pp. 303–316.

Ruigrok, W. and R. van Tulder (1995) *The logic of international restructuring* (London: Routledge).

Sachs, J.D. (1992) 'The Economic Transformation of Eastern Europe: the Case of Poland', *Economics of Planning* 25, pp. 5–19.

Sachwald, F. (2008) 'Location choices within global innovation networks: the case of Europe', *Journal of Technology Transfer* 33/4, pp. 364–378.

Sapienza, E. (2009) *FDI and Growth in Central and Southern Eastern Europe*, Quaderni DSEMS 12–2009, Dipartimento di Scienze Economiche, Matematiche e Statistiche, Università di Foggia.

Schoors, K. and B. van der Tool (2002) *Foreign Direct investment spillovers within and between sectors: Evidence from Hungarian data*, Working Paper 157, University of Ghent.

Schröder, C. (2011) Produktivität und Lohnstückkosten der Industrie im internationalen Vergleich, *IW-Trends* 4/2011.

Schumann, C. (2006) *Financial Support Schemes for Students in Selected Countries*. EIIW Discussion Paper, Wuppertal.

Schumpeter, J.A. (1912) *Theorie der wirtschaftlichen Entwicklung* (Leipzig: Duncker & Humblot), or as Schumpeter, J.A. (1949) *Theory of Economic Development* (Cambridge, Massachusetts: Harvard University Press).

Schumpeter, J.A. (1942) *Capitalism, socialism, and democracy* (New York: Harper).

Senior Nello, S. (2002), 'Preparing for Enlargement in the European Union: The Tensions between Economic and Political Integration' *International Political Science Review* 23/3, 291–317.

Sgard, J. (2001) *Direct Foreign Investments and Productivity Growth in Hungarian Firms, 1992–1999*, William Davidson Working Paper 425, University of Michigan.

Shane, S. (1994) 'The effect of national culture on the choice between licensing and direct foreign investment', *Strategic Management Journal* 15, pp. 627–42.

Sharp, M. and M. Barz (1997) 'Multinational companies and the transfer and diffusion of new technological capabilities in Central and Eastern Europe and the former Soviet Union', in D.A. Dyker (ed.) *The Technology of Transition. Science and Technology Policies for Transition Countries.* (Budapest: Central European University Press) pp. 95–125.

Sinani, E. and K.E. Meyer (2004) 'Spillovers of Technology Transfer from FDI: The Case of Estonia', *Journal of Comparative Economics* 32/3, pp. 445–66.

Sjöholm, F. (1996) 'International transfer of knowledge: The role of international trade and geographic proximity', *Review of World Economics (Weltwirtschaftliches Archiv)* 132/1, pp. 97–115, March.

Smarzynska Javorcik, B. (2004a) 'Does Foreign Direct Investment Increase the Productivity of Domestic Firms? In Search of Spillovers through Backward Linkages', American Economic Review 94/3, pp. 605–627.

Smarzynska Javorcik, B. (2004b) 'The composition of foreign direct investment and protection of intellectual property rights: Evidence from transition economies', *European Economic Review* 48, pp. 39–62.

Smarzynska Javorcik, B.K. and M. Spatareanu (2003) To share or not to share: Does local participation matter for spillovers from foreign direct investment? World Bank Policy Research Working Paper 3118, Washington: World Bank.

Smith, P.J. (1999) 'Are Weak Patent Rights a Barrier to US Exports?', *Journal of International Economics* 48, pp. 151–77.

Smith, P.J. (2001) 'How Do Foreign Patent Rights Affect US Exports, Affiliate Sales, and Licenses?', *Journal of International Economics* 55/2, pp. 411–39.

Steffen, W. and J. Stephan (2008), 'The Role of the Human Capital and Managerial Skills in Explaining the Productivity Gaps between East and West', *Eastern European Economics* 46/6, pp. 5–24.

Stephan, J. (2003) 'Evolving Structural Patterns in the Enlarging European Division of Labour: Sectoral and Branch Specialisation and the Potentials for Closing the Productivity Gap', *IWH Sonderheft* 5/2003, Halle.

Stephan, J. (2011) 'Foreign Direct Investment in Weak Intellectual Property Rights Regimes – the Example of Post-Socialist Economies', *Post-Communist Economies* 23/1, pp. 35–53.

Stephan, J. (ed.) (2005) *Technology Transfer via Foreign Direct Investment in Central and Eastern Europe – Theory, Method of Research and Empirical Evidence*, Studies in Economic Transition (Houndsmills: Palgrave Macmillan).

Stiglitz, J.E. (2002) *Globalization and its Discontents* (New York: Norton).

Stinchcombe, A.L. (1965) 'Social structure and organizations', in J.G. March (ed.) *Handbook of Organizations* (Chicago: Rand-McNally) pp. 142–93.

Svetličič, M. (2007) Outward foreign direct investment by enterprises from Slovenia, *Transnational Corporations* 16, pp. 55–87.

Svetličič, M. and M. Rojec (1994) 'Foreign direct investment and the transformation of Central European economies', *Management International Review* 34/4, pp. 293–312.

Szalavetz, A. (2000) 'Adjustment of Hungarian engineering companies to the globalising corporate network', in Bara Z. and L. Csaba (eds), *Small Economies' Adjustment to Global Challenges* (Budapest: Aula Publishing Ltd) pp. 357–76.

Tatiana-Roxana, N. (2009) 'Competitiveness and unit labour costs in Romania', *Annals of Faculty of Economics, University of Oradea, Faculty of Economics*, 2/1, pp. 444–50.

Tavares, A.T. (2001) *Strategic Management of Multinational Networks: A Subsidiary Evolution Perspective*, Paper presented at the Sixth International Symposium on International Manufacturing, 9–11 September 2001, Global Integration (The Symposium Proceeding, Institute for Manufacturing, University of Cambridge).

Teece, D.J. (1976) *The multinational corporation and the resource cost of international technology transfer* (Cambridge (MA.): Ballinger Publishing Company).

Teece, D.J. (1986) 'Profiting from technological innovation: Implications for integration, collaboration, licensing and public policy', *Research Policy* 15, pp. 285–305.

Thompson, J. (1967) *Organizations in action* (New York: McGraw-Hill).

Todo, Y. and K. Miyamato (2002) *Knowledge Diffusion From Multinational Enterprises: The Role of Domestic And Foreign Knowledge-Enhancing Activities*, OECD Development Centre, Technical Paper 196.

Torlak, E. (2004) *Foreign Direct Investment, Technology Transfer and Productivity Growth in Transition Countries – Empirical Evidence from Panel Data*, Center of Globalization and Europeanization of the Economy Discussion Paper 26, Georg-August-University of Göttingen.

Tridico, P. (2011) *Institutions, Human Development and Economic Growth in Transition Economies* (Basingstoke: Palgrave Macmillan).

UNCTAD (1993) 'Intellectual Property Rights and Foreign Direct Investment', *Current Studies* Series A 24 (New York: United Nations).

UNCTAD (1996) *The TRIPS Agreement and Developing Countries* (New York: United Nations).

UNCTAD (2005) *World Investment Report 2005: Transnational Corporations and the Internationalization of R&D* (United Nations, New York and Geneva).

UNCTAD *World Investment Reports of various years 1999–2010* (United Nations, New York and Geneva).

van Brabant, J.M. (1980/2012) *Socialist Economic Integration – Aspects of Contemporary Economic Problems in Eastern Europe* (Cambridge: CUP).

Van den Bosch, F.A.J., H.W. Volberda, and M. De Boer (1999) 'Co-evolution of firm absorptive capacity and knowledge environment: Organizational forms and combinative capabilities', *Organization Science* 10/5, pp. 551–68.

Varblane, U., D.A. Dyker, D. Tamm, and G.N. von Tunzelmann (2007) 'Can the National Innovation Systems of the New EU Member States Be Improved?' *Post-Communist Economies* 19/4, pp. 399–416.

Veblen, T. (1915) *Imperial Germany and the industrial revolution* (London: Macmillan).

Venkataraman, S. and A.H. Van de Ven (1998) 'Hostile environmental jolts, transaction set, and new Business', *Journal of Business Venturing* 13/3, pp. 231–55.

Vernon, R. (1966) 'International investment and international trade in the product cycle' *Quarterly Journal of Economics* 80, pp. 190–207.

Veugelers, R. and B. Cassiman (2004) 'Foreign subsidiaries as channels of international technology diffusion: some direct firm level evidence from Belgium', *European Economic Review* 48, pp. 455–76.

von Hippel, E. (1976) 'The Dominant role of users in the scientific instrument innovation process', *Research Policy* 5/3, pp. 212–39.

von Hippel, E. (1978) 'Successful industrial products from customer ideas', *Journal of Marketing* 42/1, pp. 39–49.

von Hippel, E. (1988) *The sources of innovation* (New York: Oxford University Press).

von Hippel, E. (2005) *Democratizing Innovation* (Cambridge (MA) and London: MIT Press).

von Tunzelmann, G.N. (1995) *Technology and Industrial Progress – The Foundations of Economic Growth* (Cheltenham: Edward Elgar).

von Tunzelmann, G.N. (2004) 'Network alignment in the catching-up economies of Europe', in F. McGowan, S. Radosevic, and N. von Tunzelmann (eds) *The Emerging Industrial Structure of the Wider Europe* (London: Routledge) pp. 23–37.

von Tunzelmann, G.N., J. Günther, K. Wilde, and B. Jindra (2010) 'Interactive Dynamic Capabilities and Regenerating the East German Innovation System', *Contributions to Political Economy* 29/1, pp. 87–110.

von Zedtwitz, M. and O. Gassmann (2002) 'Market versus technology drive in R&D internationalization: four different patterns of managing research and development', *Research Policy* 31, pp. 569–88.

Votteler, M. (2001) *Messung der Position von Regionen bei ausländischen Direktinvestitionen in Wirtschaftliche Problemstellungen im Vorfeld des EU-Beitritts* (Dresden: W. Gerstenberger) pp. 141–51.

Wagner, J. (2010) 'The Research Potential of New Types of Enterprise Data based on Surveys from Official Statistics in Germany', *Schmollers Jahrbuch – Journal of Applied Social Sciences Studies* 130/1, pp. 133–42.

Wang, J.-Y. and M. Blomström (1992) 'Foreign Investment and Technology Transfer: A Simple Model', *European Economic Review* 36/1, pp. 137–155.

Welfens, P.J.J. (1995) *Die Europäische Union und die mittelosteuropäischen Länder: Entwicklungen, Probleme, politische Optionen*, Bericht des Bundesinstitut für ostwissenschaftliche und internationale Studien, Köln, 7–1995.

Welfens, P.J.J. (2011) 'Growth and structural change in Europe and China: new perspectives', in Y. Ma, M. Taube, and D. Cassel (eds) *Challenges to Long-run Economic Growth in Europe and China* (Marburg: Metropolis).

Welfens, P.J.J. and A. Zwiatek-Kubiak (eds) (2005) *Structural Change and Exchange Rate Dynamics – The Economics of EU Eastern Enlargement* (Berlin: Springer).

Welfens, P.J.J. and D. Borbély (2009) *EU-Osterweiterung, IKT und Strukturwandel*, Band 4 of Europäische Integrationund Digitale Wirtschaft (Stuttgart: Lucius Lucius).

Welfens, P.J.J. and P. Jasinski (1994) *Privatisation and Foreign Direct Investment in Transforming Economies* (Aldershot: Dartmouth).

Wells, L.T. (1983) *Third World Multinationals* (Cambridge, MA: MIT Press).

Westney, D. E. (1994) 'Institutionalization theory and the multinational corporation', in S. Ghoshal and D. E. Westney (eds) *Organization Theory and the Multinational Corporation* (New York: St Martin's Press) pp. 53–76.

White, H. (1982) 'Maximum Likelihood Estimation of Misspecified Models', *Econometrica* 1, pp. 1–26.

White, R. and T. Poynter (1984) 'Strategies for foreign owned subsidiaries in Canada', *Business Quarterly* (Summer), pp. 59–69.

Williamson, J. (1989) 'What Washington Means by Policy Reform', in J. Williamson (ed.), *Latin American Readjustment: How Much has Happened* (Washington: Institute for International Economics).

Williamson, J. (1993) 'Development and the "Washington Consensus"', *World Development* 21, pp. 1239–336.

Williamson, O.E. (1975) *Markets and Hierarchies: Analysis and Anti-trust Implications* (New York: Free Press).

Williamson, O.E. (1981) 'The modern corporation: origins, evolution, attributes', *Journal of Economic Literature* 19, pp. 1537–68.

Wooldridge, J.M. (2002) *Econometric Analysis of Cross Section and Panel Data* (Cambridge: MIT-Press).

Wooster, R.B. and S.D. Diebel (2010) 'Productivity Spillovers from Foreign Direct Investment in Developing Countries: A Meta-Regression Analysis', *Review of Development Economics* 14/s1, pp. 640–655.

Xu, B.C. and P. Eric (2005) 'Trade, Patents and International Technology Diffusion', *Journal of International Trade and Economic Development* 14/1, pp. 115–35.

Yamin, M. and J. Otto (2004) 'Patterns of knowledge flows and MNE innovative performance', *Journal of International Management* 10, pp. 239–58.

Yan, A. and B. Gray (1994) 'Bargaining power, management control, and performance in US-China joint ventures: A Comparative Case Study', *The Academy of Management Journal* 37/6, pp. 1478–517.

Young, A. (1928) 'Increasing Returns and Economic Progress', *The Economic Journal* 38/152, pp. 527–42.

Young, S., N. Hood, and S. Dunlop (1988) 'Global strategies, multinational subsidiary roles and economic impact in Scotland', *Regional Studies* 22/6, pp. 487–97.

Zander I. (1994) *The Tortoise Evolution of the Multinational Corporation: Foreign Technological Activity in Swedish Multinational Firms 1890–1990* (Stockholm: Institute of International Business).

Zanfei, A. (2000) 'Transnational firms and the changing organisation of innovative activities', *Cambridge Journal of Economics* 24, pp. 515–42.

ZEW (2007) *Innovationsverhalten der deutschen Wirtschaft, Indikatorenbericht zur Innovationserhebung 2006* (Mannheim).

Zhao, M. (2004) *Doing R&D in Countries with Weak IPR Protection: Can Corporate Management Substitute for Legal Institutions?*, University of Minnesota, mimeo.

Zukowska-Gagelmann, K. (2001) *Productivity Spillovers from Foreign Direct Investment. The Case of Poland* (Frankfurt/Main: Peter Lang Verlag).

Index

Note: 'n' indicates note numbers

Printed and bound by CPI Group (UK) Ltd, Croydon, CR0 4YY